T0321184

MEDICAL GENETICS CASEBOOK

CONTEMPORARY ISSUES IN BIOMEDICINE, ETHICS, AND SOCIETY

Medical Genetics Casebook, by **Colleen Clements,** *1982*
Who Decides? edited by **Nora K. Bell,** *1982*
Birth Control and Controlling Birth, edited by **Helen B. Holmes, Betty B. Hoskins, and Michael Gross,** *1980*
The Custom-Made Child?, edited by **Helen B. Holmes, Betty B. Hoskins, and Michael Gross,** *1981*
Medical Responsibility, edited by **Wade L. Robison and Michael Pritchard,** *1979*
Contemporary Issues in Biomedical Ethics, edited by **John W. Davis, Barry Hoffmaster, and Sarah Shorten,** *1979*

MEDICAL GENETICS CASEBOOK

A Clinical Introduction to
Medical Ethics Systems Theory

by

Colleen D. Clements

University of Rochester Medical Center
Rochester, New York

The HUMANA Press Inc. • Clifton, New Jersey

Library of Congress Cataloging in Publication Data

Main entry under title:
Clements, Colleen D.
  Medical genetics casebook.
  (Contemporary issues in biomedicine, ethics, and society)
  Includes index.
  1. Medical ethics—Case studies. 2. Prenatal diagnosis—Moral and ethical aspects—Case stuides. 3. Medical genetics—Moral and ethical aspects—Case studies. I. Title. II. Series.
[DNLM: 1. Ethics, Medical—Case studies. 2. Prenatal diagnosis—Case studies. 3. Eugenics—Case studies. 4. Abortion, Induced—Psychology—Case studies. W 50 C626m]
R724.C525   1982   174.2   81-82200
ISBN 0-89603-033-4   AACR2

© 1982, The HUMANA Press Inc.
Crescent Manor
PO Box 2148
Clifton, NJ 07015

Printed in the United States of America.

PREFACE

The Direction of Medical Ethics

The direction bioethics, and specifically medical ethics, will take in the next few years will be crucial. It is an emerging specialty that has attempted a great deal, that has many differing agendas, and that has its own identity crisis. Is it a subspecialty of clinical medicine? Is it a medical reform movement? Is it a consumer protection movement? Is it a branch of professional ethics? Is it a rationale for legal decisions and agency regulations? Is it something physicians and ethical theorists do constructively together? Or is it a morally concentrated attack on high technology, with the practitioners of scientific medicine and the medical ethicists in an adversarial role?

Is it a conservative endeavor, exhibiting a Frankenstein syndrome in Medical Genetics ("this time, they have gone too far"), or a Clockwork Orange syndrome in Psychotherapy ("we have methods to make you talk-walk-cry-kill")? Or does it suffer the affliction of overdependency on the informal fallacy of the Slippery Slope ("one step down this hill and we will never be able to stop") that remains an informal fallacy no matter how frequently it's used? Is it a restricted endeavor of analytic philosophy: what is the meaning of "disease," how is "justice" used in the allocation of medical resources, what constitutes "informed" or "consent?" Is it applied ethics, leading in clinical practice to some recommendation for therapeutic or preventive action?

This incomplete list of questions indicates just how complex, sometimes inconsistent, and certainly still in ferment the field of medical ethics is currently. There are no easy answers to any of these questions, nor easy choices among the alternatives. Reflexively, medical ethics seems to follow its most common method for analyzing medical issues: the study of dilemmas or hard choices. And so, faced with this overarching problem of the nature of medical ethics, we can ask a few general questions ourselves:

1. Have we asked the wrong questions, paid attention to the wrong problems, or set up polar choices in which either choice is unsatisfactory?

2. Are we still repeating the subjective vs objective controversy with some rephrasing: Is it the primacy of science or

values, of technology or patient concerns, of medical advances over humanistic issues?

3. Is medical ethics a subspecialty of ethics, of medicine, or both? Are we primarily concerned with developing our ethical tools, or with practicing good clinical medicine, or with developing some combination of some of these skills? And to what purpose?

This last question is a central one for the field. Its answer is not at all clear. Clinicians, by training and experience, would almost automatically answer: the welfare of the patient. The provision of a definition for the responsibility, the accountability, and the compassion of that answer may need to be the cornerstone of medical ethics endeavors. It is such an assumed ethos of medical practice that physicians in fact may fail to see the need to even articulate it. It simply informs everything they do. And it serves as a criterion to evaluate any specialty's contribution to medical practice: how can it benefit work with patients. Yet my sense is that this ingrained answer, reflecting the real value structure of medicine, is one of many competing replies in the field of medical ethics, again almost none articulated. Trying to make explicit the purpose of medical ethics is *not* an easy task, once it is actually begun. That was made exceedingly clear during a recent interview at a large medical center. The conversation involved the possible interaction of a medical ethicist with the Psychiatry Department. The clinician asked the ethicist: "What can you do for me?" The abstract answer about conceptual clarity, applied ethics, problem identification, and skill in handling the issues current in the literature sounded intellectually fine, but felt hollow to this respondent. It *was* hollow. The physician was asking how ethical analysis could benefit his patient, could enrich his practice for his patient's good. The answer to that, from my perspective, is that medical ethics can give voice to the choices, goals, and standards of behavior in medicine, can make clear that good case management is good ethical choice, and through that articulation of the values of scientific medicine, contribute to proper case management. But even this is only one perspective in medical ethics, one that sees the discipline as an interdisciplinary endeavor primarily concerned with scientific value systems and the development of a modern ethical theory to explain those systems, though always with the goal of benefiting individual human beings. The alternatives to this perspective also need development. I can only suggest that such perspectives may involve concerns with social good (the apportionment of resources, the transfer of power), the upholding of some extra-human principles (the search for a just order, e.g.), or the refinement of terminology and the sharpening of analytic tools. I leave to the reader, and to the future of medical ethics, the problem of determining which of these alternative purposes can be the most relevant and fruitful. My own choice is clear: that the goal or purpose of medical ethics is clinical, for the patient's welfare, and is designed to express the value system of medical practice; that clinical case management is ethics; that medicine is applied ethics.

* * *

It has been a long while now since the split between the clinician and the theoretician was first overtly acknowledged as a problem in medical ethics. Some perception of the seriousness of this cleft still exists, and there is a pervasive feeling that it is now quite detrimental and should be bridged. But this may often be no more than a ritual piety, since it is clear that the implementation of a program of unification would actually carry with it major consequences that those now working in the field of medical ethics may not want to accept. In fact, without significant changes in the attitudes and conceptual structures prevalent in each of the contributory disciplines, such a healing interdisciplinary program may not be viable, and the ethicist may well be engaged in no more than rhetorical posturing.

Ethicists trained in the humanities tend to approach medical ethics from traditional structures that may or may not reflect the current realities of medical practice—and I would contend most often do not—and generally make no provisions for the incorporation into their theories of self-correcting feedback from medical experience. Ethical structures are generally imposed, without opportunity for change and growth, as predetermined categories for dealing with that experience. And although such structures (if flexible and adaptive) can supply powerful generalizing functions, their empirical data base is often missing, out of date, or unable to exert modifying effects on these rigid ethical theories. It is always assumed that we can apply these extant and unchangeable (or unaffected) theories to new experiences, and that the experiences themselves can be explained and evaluated under one of the theory's prior classifications. An appreciation of the concept of interaction, except by Alasdair MacIntyre and some few others, is quite lacking. The concept of change is often regarded as a "scientism" (in literature, the *Brave New World* model) and as clearly "inhumane" or nonhuman. Although there is always a tension between stasis and breakdown, between accepted theoretical structures and new configurations, we need to excercise special care to avoid the trap in which such static patterns of thought overwhelm future possibilities or options. To be useful in the real world, categories should be devised in such a way as to maintain sufficient flexibility to allow treatment of all options, and certainly such categories should not arbitrarily restrict options. And those trained in the humanities are at particular risk of falling into this trap unless their theoretical structures soon develop better internal monitoring and self-corrective mechanisms.

It is precisely here that medicine—because it is a clinical, and therefore a continually self-corrective, process—can be so valuable to ethics. But to be of use, the following requirements must be satisfied by any remotely responsive ethical structure:

(1) An actual understanding of current practice and knowledge (the state of the art and science) must obtain. This must be a real understanding—not a predetermined framework applied to the field, but an actual awareness of what is in fact going on. First-hand experience, the actual transaction between the observer and the environment, should not be devalued; it is unique and probably incapable of substitution. Feedback in medical practice is usually instantaneous and unmediated. Such direct feedback, in which there are no intervening steps, allows less room for distortion (noise), is a more reliable check with experience. Although there is always some garbling of signal (message), the longer the process continues, the more difficult it is to distinguish the signal from the noise. Failure to meet this requirement of a real understanding of medicine leads to distortion, as the general level of ethical comment on Behavior Therapy, for example, illustrates. Few practicing Behavior Therapists would recognize their discipline reading that critical literature, which tends to fall into *Clockwork Orange* preconceptions. The significant philosophic danger is the perennial straw man, transplanted to medical considerations.

(2) The recognition must be developed, and acknowledged, that empirical events can necessitate change in our operating ethical theories. I am not here arguing for "pure" empirical events, but neither would I admit a homogeneous theoretical interpretation for such events. Since we are bombarded with alternative experiences, when enough of them are out of line with theoretical frameworks, it seems clear that the frameworks must either be extensively modified (in systems language, compensating mechanisms must come into play to maintain stasis), or they must be abandoned (systems collapse). Although the former is less

traumatizing, the latter is frequently more fruitful in the long run, and there is no good philosophic argument for preferring modification to abandonment of a theory. For a somewhat different formulation of this thought, I would recommend study of Willard VanOrman Quine's approach in *From a Logical Point of View*. The decision to modify or abandon a theory finally becomes a matter of the complexity of the compensating mechanisms, and the energy cost and limits of these mechanisms.

If these two requirements are met, the clinical perspective can effectively modify—and even generate—the theoretical perspective. I would argue that, in a new field, the emphasis should go in that direction, rather than from the theoretical to the clinical. In any event, there is already sufficient emphasis on the theory-to-practice direction, and a redress of the balance would be a small goal.

Medical ethics, then, needs to be an interdisciplinary field in a particular sense of interdisciplinary. Although it is easy to get agreement on the need to be interdisciplinary, it is not so easy to get agreement on what we mean by that. There are more ways than one to be interdisciplinary, leading to very different results and very different sorts of medical ethics. I am using the term in the sense of the interdisciplinary sciences: biochemistry and molecular genetics, for example. The assumptions and methodologies of both fields are open to question and modification by each other. Because of the interaction, both fields can fundamentally change. Those working within the interdiscipline need adequate and first-hand knowledge of both fields. "Two culture" thinking needs to be given up: not the culture of medicine and the culture of ethics, but a science-ethics.

Ideally, then, this book should serve three functions:

As a source book of case studies in medical ethics, most specifically in the area of prenatal detection.

As an outline of the ethical problems and areas of concern generated by a working prenatal detection program, and thus a blueprint for more extensive work.

As a possible working model for the discipline of medical ethics. Our scope is by necessity extensive rather than intensive, but again relying on the example provided by systems theory, some levels are best handled by extensive methods rather than intensive ones.

Those chiefly interested in the theoretical aspects of medical ethics—or concerned primarily with the technical philosophical arguments—should consider beginning with the last chapter of this book, rather than the first. In Chapter 8, a systemization of the concepts that were developed during the course of a three-year data collection period is carefully elaborated. The structure of the book, then, can be seen to emerge from the interaction of daily clinical experience with theoretical ethical hypotheses, an interaction that fosters the gradual development of a powerful medical ethics systems theory grounded in real clinical data. As a result, the philosophically oriented reader primarily interested in medical ethics systems theory and its justification, rather than in its case history development per se, may find it more rewarding to start with the summarizing Chapter 8.

For the above reasons, this work adopts a Case Study approach, based on my involvement in the working of an actual Prenatal Detection Program affiliated with a University Medical Center. The identification of ethical issues arises from the actual cases under discussion, and the various alternative interpretations must, and do, always refer continuously back to that case, or to others in the same class. This is not to presuppose an assumptionless stance, because humans are socialized animals and are therefore raised on assumptions, and any theory must automatically incorporate a certain number of them. But again, since our conditioning environment is hardly monolithic, each individual represents a certain selective variation. In fact, within one individual, the selected primitive assumptions are also hardly monolithic, or we would not have conflicting internal needs and interests. It is precisely this theoretical plurality that best equips us for ethical adaptation in the light of continuing experiences, as realized in a clinical perspective. Such an approach rather nicely models the evolutionary and ecological perspectives of the life sciences. At this point, the reader may realize that we have reintroduced the theoretical perspective as well. I am not calling for a dichotomy—the soul and joy of some philosophy—but a transaction (to speak Pragmatically) or a stasis (to speak systems theoretically).

* * *

There are a number of problems generated by prenatal detection programs that have not previously been identified and discussed in a medical ethics framework. Since these programs show some signs that they will continually initiate future social and ethical choices or options, I have come to feel that a first-hand analysis may be of some current and future use, and may possibly help avoid confusion about and unfruitful discussions of the real issues. The current attack on amniocentesis, for example, which characterizes it as a "search and destroy" program, or which emphasizes the unknown risk of any intervention, demonstrates that those who oppose all application of the technique have not taken the time or effort to find out what it really does, what its actual performance involves, but are receiving distorted "information" removed from its natural context. Members of pressure groups are not alone in this, however. The record of philosophers on issues like PKU screening and genetic intervention in general is not always better. One of the crucially damaging assumptions of the theoretician or cognitivist is that meaningful discussions of real problems can be conducted using only one's cognitive tools and a smattering of the particular field's jargon. On the other hand, a systems theorist knows that no matter how powerful the computer, how elegant the software, if the data fed in are inadequate or inaccurate, what comes out will have no value.

For this reason, also, I feel that the present book embodies—not a totally unique, but a rare situation—one that I hope will become increasingly common: that of philosopher actually engaged in the everyday activities of a working program in a medical center. John Ladd's American Philosophical Association Committee has certainly broken ground in this direction, but there are some caveats. Meaningful experience continues to demand a transaction between the environment and the human being in it; the philosopher must be open to a new learning experience and not unnecessarily encumbered by traditional patterns of thought. I have found that work at the interface of two disciplines requires great flexibility and imagination, and I hope I have been able to bring into play the requisite creativity for a fruitful interaction.

*   *   *

The book's first seven chapters are not arranged in any particular order, but do build a new approach to medical ethics that is summarized in the last chapter. The first chapter introduces some of the questions involved in any generalized overview, questions concerned finally with what medicine in general should be all about. The remaining chapters are much more specific and reflect cases occuring in our Prenatal Detection Program.

*   *   *

Writing the book would not have been a successful venture without the supportive help of a number of persons who, though not always agreeing with my conclusions, offered much time, leads, critiques, and encouragement. Richard Doherty, MD has been a continuing warm and human medical guide and has supplied priceless opportunities for my interacting as a philosopher with the medical community. His weekly Genetics Clinic Conferences are a constant source of learning. The staff of the Prenatal Detection Program at the University of Rochester Medical Center also provided an invaluable support system: Jennifer Newhouse, Judy Garza, Catherine Donaldson, Doris Zallen. The Genetics Division of the Center continues to supply data: James Cupp, Peter Rowley, MD, and Marvin Miller, MD. Other University of Rochester Medical School departments deserve special mention. Certainly Andrew Sorensen deserves credit. The list also includes aid from John Romano, MD, Haroutun Babigian, MD, William Greene, MD, David H. Smith, MD, Frank Young, MD, William Drucker, MD. And finally, I do not want to forget the philosophers: Marvin Farber for career encouragement, Paul Weirich, Harmon Holcomb, Henry Kyburg, Lewis White Beck, Tad Clements, and John Ladd. Special thanks should also go to my good friend, Mary Ann Dicker, and my patient children, all seven of them.

Colleen D. Clements

Dedication

To Norman J. Pointer, MD who really does make things possible.

TABLE OF CONTENTS

Preface: *The Direction of Medical Ethics*   v

Chapter 1   (Cases 1–7)

Introduction to the Problem: *Is an Amniocentesis Program Good Medicine or Social Engineering?*   1

    Curative and Preventive Labels in Medicine   2
    The Unfragmented Patient   11
    Cost/Benefit Uses and Abuses   14
    We Can Never Do Only One Thing: The Ecological Model   21
    Attitudes Toward Pain and Suffering   26

Chapter 2   (Cases 8–23)

Working with Information   31

    Lack of Documentation, Verified Diagnosis, or Autopsy   32
    One Syndrome or Separate Problems   42
    Catch-All Explanations   50
    New Syndromes   54

Chapter 3   (Cases 24–51)

Denial and Reality Testing   58

    How Much Patient Denial Can Be Accepted?   59
    Anxiety Induction and Lowering   66
    Perceptions of Risk   73
    Overmedicalization or Good Medical Practice?   83

Chapter 4   (Cases 52–74)

Self-Image   87

    The Worth of the Individual   87
    Carrier Status   89
    Affected Parents   96
    Affected Offspring   103
    Parents' Sense of Responsibility for the Defect   109

Chapter 5   (Cases 75–92)

Experimental Research and Procedures   114

Thalassemia Research   115
Other Applications of Fetoscopy   121
HLA Linkage, Secretor Linkage Research   125
Blood Sampling Studies, Amniotic Fluid Sampling Studies,
  and Psychological Studies   130
Future Effects of Environmental Contaminants   135

Chapter 6   (Cases 93–118)

Selective Abortion: *Range of Choices and Time Frames*   138

Counseling Decision After Physician Referral   145
Prenatal Testing Decisions   147
Abortion Decided Affirmatively or Negatively Before Testing   151
Abortion Decided Affirmatively or Negatively with Test Re-
  sults   154
Followup Support for Abortion Decisions   158

Chapter 7   (Cases 119–130)

Social and Individual Interest Conflicts   165

Sex Determination and Selection   168
Excuse for Aborting Unwanted Pregnancies   176
Referral and Feedback   178
The Physician in the Dual Role of Filling Social Interests and
  Individual Interests   182

Chapter 8

Pragmatics as an Ethical System   190

Adaptation and the Pragmatic Criterion   190
The Traditional Trio in Medical Ethics   192
Pragmatics   208
Value of a Systems Model   212
The Nature and Justification of Intervention   217
The Forgotten Human Condition   219
The Worth of Human Beings   222
Future Issues in Medical Ethics   223
Medicine as Mythology   227

Index of Case Studies   229

Index   231

MEDICAL GENETICS CASEBOOK

Chapter 1

Introduction to
the Problem

Is An Amniocentesis Program
Good Medicine or
Social Engineering?

Although some would not see it as totally heretical to suggest that intellectual discourse has its fashions, few of us are eager to consider the possibility that even our rational enterprises are at best partially, and at worst completely, the solidified product of our lifelong conditioning. The general fashion in medical ethics discussions, however, is easy enough to pick up:

(1) Too many nonmedical problems have been forced into the medical model; we have over-medicalized our existence. There is a proper, and much narrower, task for medicine and the profession and its practice should be confined within those limits.

(2) Medical information and intervention restricts our freedom; it is assumed that in our daily life we actually enjoy such a pure freedom, or that it is theoretically attainable.

(3) Medicine is political and can have serious negative effects on social well-being. Therefore, medicine needs to be politicized openly by involving consumer and government agencies in its future development and regulation.

(4) The current health care system must be fundamentally changed.

(5) The government should enlarge its regulatory function and directly limit research and application of research.

(6) Physicians are technocrats.

(7) The Rights Model is the ethical model of choice.

There are serious problems with accepting any of these assumptions, and in the course of this book I hope to challenge their reflex adoption.

1

Curative and Preventive
Labels in Medicine

Case Study 1: Newborn
Sickle Cell Anemia Screening

The initial goal of this case study, and of those that will follow throughout the book, is to identify in specific context some of the key ethical questions or issues we confront today in medicine, most particularly in our genetic counseling and prenatal detection programs. The general discussion following the case studies in each section, and in the several chapters, will build toward a general ethical system, based on a detailed consideration of these questions, that will leave us equipped to answer them or to give the road map toward such answers. The first task, however, is to ask the right questions.

There is considerable morbidity during the first 6 months of life for infants suffering Sickle Cell Anemia (SCA). Meningitis, pneumonia, and other infections that take hold before the infant's blood disorder is diagnosed are a particular problem. If screening for the syndrome were mandatory for all newborns at special risk, this would be highly preventive by assuring the use of prophylaxis (penicillin, Influenza H vaccine). However, in rare cases, the test can result in false positives. In one such instance, twins were incorrectly given false-positive results, with subsequent serious psychological problems for the mother, who had become afraid to handle or interact normally with her babies. The issue of how a mother's perception of her child can be altered by medical information is a problem of education, long-term supportive physician interaction, and defining the advantages and disadvantages of providing such information. The problem of false positives on this particular screening test can, however, readily be handled by also screening the maternal serum for all positive results. Although an additional expense, and one that benefits only a very small percentage of the total screened population, we need to ask whether the social cost of such expensive secondary screening should ethically override the individual cost that arises when lack of screening damages the mental health of those afflicted or their families. What is the actual cost? In this case, after approximately 6 months, the mother correctly observed that the babies seemed normal and requested verification of the diagnosis. The error was then corrected. We must ask ourselves now whether the possibility of very infrequent error should block the population at risk from the real preventive benefits of this screening? Can medical interaction provide effective ways to deal with the effects of information? It is well to remember that our actions always have multiple consequences, in this case, preventive measures involve an early intervention in the SCA process, preventing infectious complications from developing. But this process is also therapeutic. And it naturally has a price.

Prenatal detection programs everywhere seem to land directly in the same sticky web of questioning the proper aim and definition of medicine on which the entire medical enterprise is now caught. Such re-examination usually signifies a loss of confidence in what we *really* are doing and of our embarcation on a search for a new normative measure, a value judgment concerning what we *should* be doing. One major conceptual division, often departmentalized in medical schools, is that between the curative (therapeutic) and the preventive aspects of medicine. Another suggested distinction is Kass' dichotomy between health and happiness (adjustment).[1] Both have particular significance for prenatal detection programs, and represent issues that must be examined and settled before we can proceed to problems more specific to such programs.

Historically, there have always been preventive aspects of medicine, even if such aspects were co-mingled with mythic and superstitious rituals (talismans to ward off evil spirits, purging to purify and prevent illness, herbal contraceptives, a host of other purification rituals, bleeding, and quarantine). But clearly, our ability in this area has expanded somewhat, even if therapeutic reality does continue to fall short of our hopes and pretensions. Vaccination is the clear and major model of new preventive possibilities. Cholesterol diets, clean air and anti-smoking measures, fitness programs and other life-style interventions are not such clear models. Chlorination treatment technology, however, suggests that an intervention can be preventive in one perspective and highly suspect in another (the production of carcinogenic chlorocarbon compounds). It is important to remember that preventive interventions also have a price, which is one of the staples of ecology.

Prenatal detection programs can be preventive medicine in two senses: (1) through carrier detection, identified heterozygotes

---

[1]Leon Kass, "Regarding the End of Medicine and the Pursuit of Heath," *Contemporary Issues in Bioethics*, Tom L. Beauchamp and Leroy Walters, eds., Wadsworth, Belmont, Ca., 1978, pp. 98–108.

for autosomal recessive disorders may decide not to marry another carrier or may decide to restrict their reproductive options in order not to risk a pregnancy in which the gene would be expressed; or, an affected individual with a dominant mechanism may decide not to reproduce. (2) An affected fetus may be detected *in utero* through various tests on the cellular and noncellular components of amniotic fluid, through ultrasound, amniography, or fetoscopy, and the option of an abortion may be chosen. The latter is preventive in a much looser sense than the former. After all, the genetic defect has been expressed in the fetus, and the intervention of therapeutic abortion prevents its expression on a different level of attention: not the biotic, material reduction level but the social system level, particularly the family system level. Simply, we prevent the development and birth of a defective fetus, not the conception of that fetus. Even the first sense of prevention does not give us as much control as we would like. We are still forced to make hard choices, to adapt ourselves to the reality of a defective genotype, rather than being able to modify the reality of that genotype. Here the prospects of genetic engineering enter—the possibility of correcting the defective genotype itself—thereby rendering such hard choices no longer necessary. But are we now talking about preventive or curative medicine? The distinction, like most distinctions, is beginning to collapse. And since, in a dynamic system, cures have a tendency not to be permanent, what we have been calling preventive medicine again comes into the picture. In our specific example, given that the genotype has been manipulated and now conforms to a functioning norm, and since it is a system in time, a naturally occurring mutation rate or a mutagenic agent introduced into the environment can, in a large enough sample, reintroduce the old defect or create a new one. Knowing some of the agents in this process, we can attempt to take preventive measures, but at this point we need to ask just what we are trying to say with these terms. It may be nothing more complicated than this: When the medical field has accumulated enough knowledge of certain physiological and disease dynamics, it can anticipate health consequences and point to etiologies. If we have even further information, we can use that model to determine how best to intervene at the etiological level, to change actualities, and to prevent the progression to the conceptual entity we call disease. This preventive function of medicine is a matter of information gathering and utilization at an early point in some homeostatic process, rather than at a later point when the disease is labeled and we call intervention curative. It occurs at either the leading edge of basic research or as a result of the attention and interest we wish to give to an early point in some physiological process. If, however, we are lacking sufficient information to predict the consequences or identify the etiology, or if our attention and interest focuses on a later point in that disease process, then our intervention attempts to block its further development and to reinstate as best we can some previous metabolic balance, and this is the curative (therapeutic) aspect of medicine. Disease, like health, is a process, a functioning.

The labels preventive and curative, then, are just convenient markers for that point at which we undertake intervention, or for the nature of the attention and interests of the physician considering intervention. They do not represent two different kinds of medicine. Nor does preventive medicine constitute a new aim or role for medicine. Medicine has always been interventionist, by definition. Whether one intervenes sooner or later is a matter of ability and purpose rather than philosophical distinction.

There is no reason, then, in terms of analysis, to restrict the medical role to that of curative intervention. Such a restriction is sometimes proposed, however, in the name of preventing medicalization of our various life problems.[2] This may be a rather conservative move, at the base of which is the difficulty of adjustment to rapidly increasing quantities of medical information. It requires much more adjustment of the social system to monitor and eliminate mutagens from the environment than to build more facilities to care for a small rise in birth defects. It will be a disruptive process to fashion mechanisms for handling mutagens. In some respects, it is easier to suffer and die than to live.

A prenatal detection program should not be embarrassed, then, to find itself functioning essentially in a preventive role. This merely confirms that such programs are based on newly acquired knowledge that focuses our attention on ever earlier moments in the development of malfunctions, and as a result expands our options beyond what was previously possible. Nor is this medicalization of such choices anything more sinister than an increase in available information, one that permits a better understanding of what our choices really are. I intend to discuss information as a value in a later chapter, specifically in the context of conflict with other values, such as anxiety reduction, and more generally, with

[2]For a brief review, see Renee Fox, "The Medicalization and Demedicalization of American Society," *Doing Better and Feeling Worse*, John H. Knowles, MD., ed., Norton, Inc., New York, 1977, pp. 9–22.

respect to what basis there may be for such valuations at all. For now, I merely wish to point out the basic structure of the over-medicalization argument, or what I think its underlying structure may be. Whether information should ever be restricted and its growth curtailed, and whether intervention can be justified, are problems I plan to deal with in detail.

Basically, the same analysis used for the preventive/curative distinction can be used for Kass' health/happiness (adjustment) distinction, with the additional consideration of his implied split between physiological functioning and psychological functioning. Explicit in Kass' rejection of behavior modification as within the province of medicine (since he presumes its goal is social adjustment), it is implicit in his rejection of happiness or contentment as a concern of medicine, and his entirely physiological and overtly behavioral description of "well-functioning." Health, he feels, is a natural norm and the only proper concern of medicine, but this concern is strictly limited since it must not alter humans (psychologically or genetically) without transcending the province of medicine. This peculiarly static view of human nature—that it is or should be fixed or immutable—is again the conservative response to the increase in medical information and to the expanding options that generate larger oscillations in the dynamic system that medicine comprises. Kass' view is combined with the physiological/ psychological split in which human feelings and purposes are removed from the domain of medicine to lodge, one imagines, in some moral domain where moral norms are operative, rather than the "natural" norm of medicine. As a systems theorist, I am puzzled how one can discuss well-functioning and manage to leave out goals, purposes, interests, and feelings. I have little doubt that all the content of this level (goals, purposes, happiness, social adjustment, well-functioning) could in theory be reduced to the level of a biochemical system having functional norms, although such a reduction would probably not prove fruitful. The important point is that *all* of this level could be so reduced—the *whole* organism as a functioning physiological and psychological unit can be viewed as a biochemical system. There is no good theoretical reason for medicine to be more concerned with any one aspect of this functioning than another, although practically, its interests and successes certainly vary. If one is actually concerned with the whole organism, one is concerned with purpose and motivation, and indeed with the full range of medical interventions that Kass wishes to exclude as not aimed at health care, but rather at satisfying patients' wishes (artificial insemination, cosmetic surgery, vasectomies, abortions, unspecified psychiatric interventions, etc.). One is also then concerned with the proper functioning of the social species humans constitute, which leaves Kass no reason to exclude various behavioral therapies from medicine as social adjustment or moral virtue goals. (A reading of the *Annual Review of Behavior Therapy* or of the writings of David Begelman, however, would clearly show that social adjustment is not the single, or often even the most important, aim of behavior therapy or psychoactive drug intervention.[3])

In any case, Kass really cannot have it both ways. If he wishes to use the systems model of the whole functioning organism, then all the functions of the organism must be included and certainly the goal, aim, and end-state are crucial components of an organism in systems theory.[4] Unless we accept a suppressed premise that the goal or aim is not crucial to the labeling of a norm of functioning, his argument will not work. Such a premise is clearly false in systems theory since: (1) a functioning norm is determined by the goal or aim of the organism and (2) function itself is seen in terms of a homeostatic aim. The only other way Kass' system might operate is to accept a suppressed premise that psychological and social functioning is different in kind from some sort of "pure" physiological function. I see no compelling reason to grant this premise, which would require a host of mysterious entities or disconnected functional levels to make it work. It is, after all is said, the body system that falls, runs, cries, and plans, though for convenience we use different levels of description. It would be, if we could someday manage the complexity, a waste of time and energy to describe running in biophysical terms, much less planning.

Since I am going to be using the systems model (or paradigm, more properly) throughout my book, it might be a good idea to here quickly clarify some of the basic concepts this will involve. I have already given an example of the dynamic, homeostatic concept in systems theory. An ongoing system attempts to remain in a controlled state, which involves mechanisms for maintaining swings or oscillations within certain limits sufficient to prevent such wild gyrations that the whole system breaks down. Feedback loops usually accomplish this. They signal to components that lev-

[3]*The Annual Review of Behavior Therapy*, Franks and Wilson, eds., Mazel, New York, 1973–78. David Begelman, "Homosexuality and the Ethics of Behavioral Intervention," *Journal of Homosexuality*, **2**, 213–19 (1977).

[4]*Unity Through Diversity*, William Gray and Nicholas Rizzo, eds., Vols. I & II, Gordon & Breach, New York, 1973. Ludwig von Bertalanffy, *General Systems Theory: Foundations, Development, Applications*, George Braziller, New York, 1968.

An Example of Hierarchical Levels

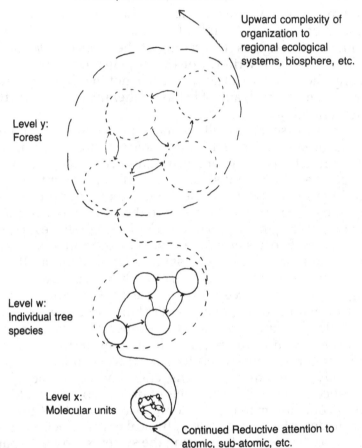

Upward complexity of organization to regional ecological systems, biosphere, etc.

Level y:
Forest

Level w:
Individual tree species

Level x:
Molecular units

Continued Reductive attention to atomic, sub-atomic, etc.

els of certain other components are changing, and this triggers a counter reaction within the system. If the loop is a negative feedback one, a rising level of one element will cause other levels to change and dampen the rise. Usually, there is a bit of overcorrection that results in a wave-like approach to equilibrium rather than a steady line of balance. Overcorrection is built-in because some finite time is always required to transmit the message about the direction of the changing level, activate the proper dampening response, then get back the message that the level change has reversed that will turn off the response. Stasis thus involves an oscillation rather than a level line.

Positive feedback loops are also dynamic mechanisms, but much more risky ones to count on if system equilibrium is to be maintained. These loops report a rising level to units of the system, and the response is to reinforce the rise. Eventually, external limiting factors bring a halt to the continually rising level (starvation in animal populations, e.g.), but this boom and bust balancing act is very precarious and can result in such a catastrophic bust that the system cannot recover, totally breaks down.

A second major concept, which will be frequently used in this book because it has great promise for the unmuddling of several philosophical muddles, is that of Hierarchy Theory, or levels of complexity/configuration. Perhaps the best example to illustrate this is a bit of folkwisdom, true common sense: "He can't see the forest for the trees." Others have used the example of Chinese boxes, but I think the forest/tree analogy might be better. We can pay attention to individual trees. Our purpose can involve enumerating such individual specimens in a woods. But we could also, in terms of our interests or in terms of the level of organization or complexity, pick out units going up or down the scale. If we switch our attention from the tree units, we enter a level in which such units relate to each other and to other units in the biosphere in such a way that we have a larger, more complex unit called the forest. Because of the various interrelationships, the forest, in ecological terms, has properties or characteristics quite different from the tree level of organization: the different canopies of a forest, the watershed system, the effect of burning on new growth, the dynamics of a climax system, the niches for animal species. In theory, the one level can be translated into the other, although in a very cumbersome way. The important point about levels, however, is that terms and constructs at one level should not migrate to another level, or if they do, that warning flags should go up that we have switched discourse or attention from one level to another. Otherwise, confusion of purpose or organization can easily result. Most of the problems of ethics (e.g., abortion) arise because levels of organization are criss-crossed haphazardly. Mixing metaphors is always a problem.

In terms of existence, the trees and the forest involve the same thing, which is why many systems theorists use the Chinese Box example. Societies are individuals, forests are trees. However, the relationships that hold between the units create features requiring new terms and new concepts to handle pragmatically with any degree of success. And we can do the same thing going down the scale of levels of organization. The tree system could be viewed as the larger configuration of a number of physicochemical units, whose characteristics and explanatory terms are very different from the terms required to explain the tree system readily.

Workers in the life sciences, and lately in physics, are quite conversant with these concepts in systems theory. The humanities still tend to use the old scientific paradigm of linear cause and effect, and of billiard ball atomic units, holding to the faith that these terms and constructs are interchangeable at all levels. Pragmatically, this simply does not work out. I hope my continued use of the systems theory paradigm throughout this book will eventually result in the reader feeling comfortable with it. It does require a change of intellectual focus, however, and the use of a language that sometimes seems awkward. The payoff is that such a language helps us dissolve and clarify our understanding of some significant philosophic problems.

The rule of thumb, then, is to determine which functions appear to have a discernible impact in causing or preventing those general body malfunctions that it is in our interest to prevent. Those functions then become medically relevant. Kass' health/happiness distinction also, in that case, dissolves.

Rather than the either/or dichotomy (perhaps a version of the double-bind) that too often characterizes theoretical analysis, it would be instructive to take a look at a clinician's approach to describing the role of medicine. John Romano develops six aspects of medical intervention, in keeping with his perception of the patient as a complex, socially integrated system: (1) diagnostic and judgmental, (2) preventive, (3) therapeutic, (4) integrative, (5) investigative, (6) educational.[5] Eric Cassell presents a sensitive discussion of the personal interactions involved in the practice of medicine, with emphasis on a genuine supportive relationship.[6] In a working prenatal detection program, Richard Doherty emphasizes the functions of educational sharing, preventing, and caring. In all these approaches, the view of the patient as a complex system, eliciting a variety of possible responses, is paramount. Jules Cohen discusses the difficulties inherent in a too-narrow concept of medicine-as-technology and the medical inappropriateness of depersonalized medicine, while emphasizing the preventive features of good medical practice.[7]

[5]John Romano, "Basic Contributions to Medicine by Research in Psychiatry," *Journal of the American Medical Association* **178,** 1147–50 (Dec. 23, 1961).

[6]Eric Cassell, *The Healer's Art: A New Approach to the Doctor-Patient Relationship,* Lippincott, New York, 1976.

[7]Jules Cohen, "Has Medicine Become Depersonalized," *Rochester Review,* Spring, 1979, pp. 19–22.

The Unfragmented Patient

Case Study 2. Lepore/β-Thalassemia

The father of the young patient himself is heterozygous for both
Lepore *and* β-thalassemia, and as a result has symptoms similar to
a homozygous person for either. He is thin and has a chronically-
ill appearance that prevents him from finding work. He required
frequent transfusions until a splenectomy proved necessary, and
has a pansystolic murmur and arrhythmia of the heart. There was
one episode of congestive heart failure. Although he and his wife
discussed with a physician the possibility of having affected children, incorrect information was given and no carrier testing was
done on the wife. Their daughter was not correctly diagnosed until there was a workup at a medical center. The child actually has
the same illness as her father, and she is heterozygous for Lepore
and β-thalassemia. She receives transfusions every two to three
weeks, often from different medical personnel. One transfusion
had to be started nine times, and the father requested at clinic that
his daughter have one consistent health care provider do the
transfusions. His daughter was upset and afraid of the process,
rendering the procedure more difficult to accomplish. In addition,
it was causing the father considerable psychological distress. Simply considering the mechanics of transfusion would seem im-
proper medical management for this patient and her family. The
child's need to relate, on a long-term basis, to a health care provider is apparent. Neither the legalistic contract model or the fiduciary model (when the physician relates to the patient as passive
object) can adequately reflect the real situation. However, as the
physician's organization grows larger and more complex, and as
public management moves toward a more bureaucratic
structuring of medicine, will the required personal interaction be
possible?

Case Study 3. Huntington's Chorea

An eleven-year-old ward patient was being evaluated for the possibility of Huntington's Chorea. His father may have HC, but refuses to be examined by physicians; second-hand reports indicate
erratic behavior, memory loss, and falling. The paternal grandmother died of HC, although her son denies this. The boy's parents are divorced. His sister, in her late twenties, is reported by her
mother to have trembling, memory loss, and mood changes, but

she also refuses examination. The boy's only reported problems seem to be minor school difficulties, a "trembling spell," and abnormal EEG. However, there are six siblings, most of whom are not doing well in school. His mother is very active in an HC group. The father has remarried and has a young child. Clearly, the family dynamics here bear watching and may be crucial for proper diagnosis. This may be as crucial (or more so) as physiological workup (e.g., CAT scan). In fact, what truly is the reason for this diagnostic workup? If it is for physiological reasons only, consider what information could be obtained. We could tell the mother and son that the boy has HC if it were an unusually early onset and the tests determined it. If the tests were negative, we could tell them nothing. The boy may or may not have HC; there's little way of knowing. The physiological payoff is either the terrible certainty that the boy has HC; or the alternative inability to reassure and relieve anxiety. In the process, the boy's anxiety is probably being raised. Still, this was a clear call for help from the boy's parents. And it seems the kind of plea that medicine has always traditionally responded to. Would limiting the response make any sense? Would appreciation of the psychological interaction make the response more appropriate? The medical management of HC, I would argue, is more than disseminating the 50% risk figure and palliative treatment for those who develop the disease. Such care is long-term supportive treatment, and requires a personal physician/patient interaction. Even if such long-term care were broken into specialties (self-help group, case worker, therapist, geneticist, neurologist), an integrative health-team effort would be required, since the patient is not conveniently fragmented into those specialties.

The Kass restriction of the medical role raises a major issue in medicine, consideration of the patient as an Open System, as a whole. Michael Bayles' proposal of the "Body Mechanic" model raises the same problem.[8] Bayles' argument illustrates some common assumptions in philosophy that poorly serve medicine:

(1) That human beings in an interaction may have only external relations to each other, not internal ones. This cold, object model of humanness is fairly common in philosophy. If the only reason we had for rejecting it were Kant's imperative not to treat

[8]Michael D. Bayles, "Physicians as Body Mechanics," *Contemporary Issues in Biomedical Ethics*, Davis, Hoffmaster, Shorten, eds., Humana Clifton, NJ, 1978, pp. 167–78. See also my reply, Colleen Clements, "Physician as Body Mechanic—Patient as Scrap Metal: What's Wrong with the Analogy," *ibid*, pp. 179–85.

rational beings as means only (the epitome of an external relationship, as an object only), we might be in bad conceptual shape. After all, the only Kantian basis for this proscription against treating human beings only as objects is that violating the rational Law of Contradiction is clearly proscribed. This thin, rational perspective is too rarefied a base, I'm afraid, though it may counter Bayles' point. I think that comparative ethology, psychodynamics, and social psychology in general demonstrate that social species, including humans, have internal relations when they interact, that these are essential components of a stable social system, and that lack of them results in aberrant and destructive behavior. This is such a crucial element of human nature that infants treated as objects only, related to in only external senses, have failed to thrive and sometimes died. Infant Rhesus monkeys, deprived after birth of any object-relations, withdraw into depressive psychosis and in a natural environment would not survive. To attempt to reduce the medical interaction to such an external relationship is to restate incorrectly that the body (patient) can be simply an object to which things are done.

(2) Although medical personnel know better than this because of their empirical encounters with psychosomatic and psychogenic illnesses, the placebo effect, patients' psychological cooperation or resistance altering outcome, and so on, philosophy appears to manage this conceptually. How? Perhaps through all the confusion generated in the philosophy of action by terms like "body" and "person." In some of these interpretations, "body" and "person" can be distinct, and this is in general accomplished by confusing levels or hierarchies of discourse. "Person" is a social label we use that is equivalent on the biophysical level to "body." The problem and confusion arises when we so mix these levels that we delude ourselves into believing that a separation can in fact be achieved, that somehow we are talking about separable entities on the same level. The physician cannot interact with another living human only on the biophysical "body" level unless both doctor and patient become some sort of science-fictional colliding-atoms monstrosity. Even if this could be accomplished, it would be bad medicine, since important aspects of the disease process identifiable on the "person" level could not be handled on the "body" level, where the biochemical components are not known or manipulable.

What, then, really goes on in the physician/patient relationship? And what needs to go on? In order to eliminate internal relationships in the interaction, we would have to demonstrate (a) that psychological factors were unimportant to the development or the course of illness and its cure, and (b) that they were not major components of many physician/patient interactions. Bayles actually does implicitly adopt those contentions. In his view, it is only a physician's conceit that patients come to them for psychological reasons rather than physiological treatment and cure.

(3) Again, the split is implicit between the psychological and physiological, a persistent assumption in the humanities. The actual data simply fail to support this required premise. A recent article in *Lancet* gives a thumbnail sketch of various studies demonstrating the psychiatric component and the significance of this for the outcome of illnesses.[9] George Engel's and other's work[10] in the area of psychosomatic medicine, research showing correlations between significant life events and disease onset or correlations between personality types and disease entities—all of these results speak to the need for a systems model of medicine.

Cost/Benefit Uses and Abuses

Case Study 4. *Amniocentesis without Precommitment to Abort*

Based on the age of both husband and wife, an older couple received genetic counseling and scheduling for amniocentesis. Public funds (Medicaid) were to finance the test. The prenatal detection program here does not require that the couple give permission for a therapeutic abortion before the test. This couple was firm about rejecting abortion as an option. However, they still wanted the test to help prepare themselves for caring for a Down's Syndrome infant should the pregnancy result in that. Is this proper use of public funds? Are there medical indications for the test? Should a therapeutic abortion agreement be obtained prior to amniocentesis?

There are a number of complicated problems packed into this case. We must first decide whether financing with public funds

[9] "Psychiatric Illnesses among Medical Patients," *The Lancet*, March 3, 1979, pp. 478–79.

[10] George Engel, *Psychological Development in Health and Disease*, Saunders, Philadelphia, 1962; "The Need for a New Medical Model: a Challenge for Biomedicine, *Science*, **196**, 129 (1977); "Emotional Factors in Gastrointestinal Illness," Arthur Lindner, ed., *Excerpta Medica*, *Amsterdam*, New York, 1973.

puts this couple into a novel category in which special evaluations will be operative. Are they required to meet different standards from a Blue Cross financed couple or a private fee-for-service-couple wealthy enough to "buy" amniocentesis? Does public financing demand a more stringent accountability than private financing? There are two general answers: (1) Public financing does require a different set of standards because we are dealing with society's resources, not private resources; or (2) The same set of standards must apply to all patients in the interest of fairness or justice, and to avoid a two-track health care system.

In the wake of Club of Rome reports, government studies predicting energy resource shortages, OPEC world trade manipulations, and the demise of the dollar, we are all generally aware of the finite nature of resources and the need to adopt a limited growth scenario. This has been a profound move because the pie to be divided cannot automatically increase its size to meet increasing demand for more or bigger slices. Redistribution often becomes the only practical solution. And redistribution means social conflict and hard choices. It is becoming a zero-sum game. So the first answer assumes a social responsibility and respectability it has not historically had. We can think in terms of increasing unfairly the medical costs society must assume or of depriving some more worthwhile need by choosing a perceived-as-lesser need. There is an implied assumption that we cannot afford an equitable health care system at current levels of care. The data seems to support this conclusion as pessimistic but realistic. A recent forecast by a business information and market research company (Predicasts, 1979) projects a 40% increase in world spending on health care by 1983 and concludes that when national health insurance plans are implemented, costs go through the roof. Canada is experiencing such problems, Australia scrapped its NHI plan when health outlays tripled, Britain has reduced benefits and limited access (inducing patients not to receive medications, rationing access to certain expensive treatments like dialysis), Sweden also has reduced benefits, denying expensive operations to the elderly.

We may be approaching some tragic choices that philosophy (and philosophical ethics in particular) has generally been too optimistic or unrealistic to consider.

Certainly a standard ethical argument could be made for fairness or just distribution of medical resources, even in terms of public and private resources, since these terms are problematic. In the long run, one could argue, there are no private resources ("spaceship earth", the Tragedy of the Commons), only publicly or privately controlled resouces. All resources are limited, and conservation appeal applies to all sectors, to the private equally as to the public. The wealthy patient who pays the total cost for amniocentesis puts as much drain on health care (and other) resources as the public patient. In a limited growth (LG) model, some other need will not be met, or met less well, whether the test is publicly or privately financed. The model does not justify a two-track health care system, then, but could be used as an argument for an equally distributed system. In the LG model, accountability can be global and classless. So for our patients requesting the test, the source of funding would not limit their access to it.

However, equal distribution of a limited pie may involve limiting *everyone's* access to the test. This ethical position on fairness forgets the darker side of existence and survival. Although it certainly follows a fairness principle to say either all will be saved, or none will be, keep in mind the overloaded lifeboat and what it really means to apply that principle. Keep in mind the individual cost of limiting access and the terrible choices of the triage system. It seems to me that simply existing also involves a survival principle, and that such living may not always allow us both principles. Do we really wish to let all in the lifeboat die if all cannot be saved, or are human beings compromisers? What does compromise do to us, and how much can be compromised? Ethics has not been very instructive on these fundamental questions concerning the human condition.

Assume the ideal, however, namely that the fairness principle is not in conflict with reality and that equal access to amniocentesis is possbile. We can still question the amniocentesis procedure without a precommitment to therapeutic abortion; we can question any amniocentesis without an abortion agreement regardless of funding. Again, we have two options: (1) amniocentesis without therapeutic abortion is cost-effective, or (2) amniocentesis without therapeutic abortion is not cost-effective. The difficulty now is to determine what we mean by cost-effective. It is usually not too difficult to work with direct economic costs. With the analysis of relatively hard data, one center's prenatal detection program can project a "programmatic break-even cost" at maternal age 32, and a plus benefit from that point on. This assumes that therapeutic abortion will be a realistic option, and in fact could not be cost-effective (considering only hard data) without therapeutic abortion. The hard data includes institutional costs for Down's Syndrome individuals, loss or gain of adult earning potential, cost of testing all pregnant women from 32 years old on, and so on.

These data offer *prima facie* support for the second option, that amniocentesis is not cost-effective without precommitment to abortion. However, difficult as it is to quantify, I would argue that "soft" data are essential in this evaluation.

Here, it seems appropriate briefly to consider some of the "soft" factors in the case study, and in the program in general. As with the hard data, we assume that births of Down's syndrome individuals will occur a certain percent of the time. The considerations then become those of the psychosocial costs: Pressure on the family structure (some data is beginning to accumulate on the cost of broken family units), individual psychological stress (the cost of mental health dysfunction can include loss of income, loss of productivity, cost of support systems), specific postpartum depression, political costs of forced abortion when individuals do not wish to honor their precommitment (costs of enforcement, and of the restriction of autonomy in this area). We assume in the analysis that foreknowledge or preparation will mitigate these costs, and that individuals in need will be identified and picked up by the support systems before the crisis stage is reached. We also assume that early identification of the problem can prevent the more extreme dysfunctions, and that the support systems will indeed be effective. There is some evidence for this. James Cavanaugh's 1978 statement before the Senate Finance Committee contains a useful summary of studies documenting the utility of early intervention and support.[11].

Naturally, the cost of these support systems must also enter the analysis. It will of course remain more cost-effective to abort a Down's syndrome fetus than not. But there may be real validity in the case study couple's desire to prepare for a Down's birth, and it may be more cost-effective to do the test, if we add the psychological benefits of such preparation to our equation. These benefits, if properly quantified, can outweigh the test cost. Admittedly, it is difficult to get a handle on these data, but there is the strong possibility of serious distortion if we proceed without such considerations; our economic analysis will seem precise and quantitative, certainly, but that appearance will be deceptive. In our program, actually, we have case study followups that demonstrate on a small scale the psychological and social benefits of amniocentesis when a defective fetus was detected, but abortion was not chosen.

[11]James L. Cavanaugh, Statement of the American Psychiatric Association before the Senate Finance Committee, Aug. 18, 1978, pp. 1–9 and Appendix A, A1–A31.

I would like to consider two points in more detail. First, a precommitment to selective abortion demands decision-making before one is aware of what one's feelings will be in the actual situation. Such a decision is subject to change, and our experience is that both affirmative and negative decisions have changed to the opposite. This sets up the possibility of a very delicate situation. Will we demand of a couple that the precommitment be honored when they now reject abortion as an option? Will we force the abortion? Without enforcement, the precommitment is simply an empty form. With enforcement, it could become a nasty business. In addition, precommitment limits the groups that could benefit from and support the program. (Another soft cost would be lack of program support and consequent demands for its terminaton, or underutilization. Such costs can be very high indeed.)

Secondly, cost/benefit consideration may give the impression that the social institutions to render these evaluations and rankings routine are already in place; little could be further from the truth. In fact, our trade-offs are made very haphazardly, in a far from rational manner. Keeping strictly within the economic sphere, in fact, cost/benefit analysis implies a free market situation in which cost and benefit rankings are neither complex nor ambiguous. Actual restrictions on consumer actions and the typical unreliability or absence of adequate information or knowledge makes even hard data evaluation subject to distortion. Although a utilitarian point of view assumes the possibility of an unambiguous ethics, a look at its application tends to support a more existentialist view of ethics as very ambiguous indeed.

The case study couple's request for amniocentesis, then, even financed with public funds, should not require precommitment to abortion in order to be met. Within their own value system, within alternative value systems, and even within the value system of cost/benefit analysis, it is a reasonable request. We should take care, however, not to lock ourselves into a pseudo-precise cost/benefit valuation. Cost/benefit calculations often arbitrarily restrict consequences and minimize interrelationships. The cascading effects of the interdependence of facts in an equation are crucial. For example, an MIT study using a model that took such relationships into account showed that the discovery of new sources of energy supplies, or new techniques for extracting the supplies, resulted in increased pollution, which would be an unacceptable outcome. Population, resources, and pollution, because of their complex interrelations, produce quite unexpected results when all those connections are properly accounted for, rather than ignored.

Thus, any discussion of the social impact or justification of a prenatal detection program requires a look at the cost/benefit tool employed in setting it up. Although such analyses appear to give us quantitatively precise information, we need to remember that behind a great deal of "hard" data are "soft" constructs. Cost/benefit analysis is no exception; it is, of course, a utilitarian approach, with all the problems of the "greatest good for the greatest number" and the winking away of the tension between individual needs and interests and social needs and interests. But even granting the value of this utilitarianism, there are some peculiar primitive assumptions inherent in the cost/benefit approach. An ideal individual is postulated: someone totally autonomous in all choices, capable of ordering options into a set of preferences, independent of any cardinal ordering principles,[12] and certainly not risk-taking—in short, an asocial unit. This theoretical construct seems derived from a social and cultural environment that places a high value on autonomous free choice, rationally determined action, prudence, and utilitarian calculus. Without these initial value judgments, which arise from a particular Anglo-American ethos, the analysis could not begin. I point out the initial value-laden step because the basic valuational character of cost/benefit analysis tends to get lost in the quantitative conclusions.[13]

There are a host of implicit value judgments made in the course of generating this cost/benefit analysis:

(1) A narrowly interpreted economic perspective of society is accepted as the basis for valuation and decision-making concerning amniocentesis. The fact is that almost all studies, no matter how conservative, have concluded that, from this perspective, amniocentesis is certainly cost-effective. But the problem is the unquestioning acceptance of such economic approaches.

(2) It is often easier to estimate direct costs than benefits. Benefits are often seen as averted costs, or difficult-to-

[12]Kenneth Arrow, "Public and Private Values," *Human Values and Economic Policy,* Sidney Hook, ed., NYU Press, New York, 1967, pp. 3–5.

[13]For detailed discussion of cost/benefit analyses, see Ernest Hook and Geraldine Chambers, "Estimated Rates of Down Syndrome in Live Births by One Year Maternal Age Intervals for Mothers Aged 20–49 in a New York State Study—Implications of the Risk Figures for Genetic Counseling and Cost-Benefit Analysis of Prenatal Diagnosis Programs," Birth Defects: Original Article Series, Vol. XIII, No. 3A, 1977; Edward Guiney, "A Question of Priorities," *Journal of the Irish Medical Association,* **66** (15) (Aug. 11, 1973).

quantify factors, or quite indirect. Not all costs, either, are apparent. Protocol decisions, which are often assumptive, are continuously made.[14]

(3) Human factors aside from pricing or medical technology are eliminated as unnecessary factors.

Uncritical acceptance of a cost/benefit approach may be short-sighted. In the long run, because mortality rather than morbidity markers are currently used, almost none of medicine can be justified in terms of the cost/benefit approach. Studies showing change or lack of change in population mortality are used because, the data being easier to collect and quantify, these studies predominate. Studies showing change in morbidity (pain, handicap, severity of illness) are rare. Since broader variables so greatly influence mortality figures, and since the medical variable is not statistically significant,[15] using such markers will throw into question the cost/benefit analyses of the entire medical program. Medicine's actual justification lies in more complex factors. The number of ill people whose problem is not self-limiting, or is acute rather than chronic, and who will respond to medical intervention, is at any time a very low number in terms of the entire population. This is particularly true for Mendelian and chromosomal genetic diseases (Mendelian, 1.0%; chromosomal, 0.5%). Physicians and medical facilities are actually social forms of health insurance, however, since we as individuals are never sure who will fall into this small, ill subpopulation, and we often cannot rule out the possibility that it could be us. Health and illness are contingent. To assure our access to medical aid, we need to assure reasonable access to a large enough representative population. This is not a "right" to health care, because I would have difficulty trying to locate the trans-human legal machinery or the law-giver that could serve as the necessary foundation for such a "natural right." It is not that medical care is given because we have the "right" to it, but rather, because it is the right thing to do. It is hardly Rawl's original position (that peculiar combination of Kant and social contract), since we always have the option to be risk takers, and frequently are. Our society is not *obliged* to insure our access to medical care; it is simply one of our collective human choices. And for an entire population, such insurance may not be cost-effective. Natural selec-

[14]For a current example in another field, see Luther Carter, "How to Assess Cancer Risks," *Science* **204**, 811 (1979).

[15]Victor R. Fuchs, *Who Shall Live? Health, Economics and Social Choice*, Basic Books, New York, 1974, pp. 30–56; *Health Care: An International Study*, Robert Kohn and Kerr I. White, eds., Oxford University Press, New York, 1976, pp. 394–400.

tion is actually much more cost-effective for an entire species. Such Social Darwinism (made explicit) would enjoy little acceptance because of important arational considerations: self-love, social relationships (family ties, bonding, etc.), and compassion (the fairly uniform animal response to distress).[16]

Medical benefits tend to be of the following sorts: reduction of pain, extension of functioning, shortening of the healing process, allaying of fears and anxieties, providing human comfort during an inevitable dying process, human attention, providing hope for the future, and the sense that all is being done that could be done. Such benefits are precisely the sort that economic analysts claim are too difficult to quantify and therefore abandon in their analyses. The same considerations (quality of life) are often placed beyond the sphere of medicine-as-technology, but a quick reading of the history of medicine attests to their importance. If these benefits are not quantified and included in cost/benefit studies, the character of medicine is seriously distorted and misrepresented. The fundamental problem, then, with these economic analyses is that the actual practice of medicine requires consideration of the very factors that they are not equipped to handle.

The more general philosophic problem is how to deal with individual needs and interests, and social needs and interests—with what John Ladd would call micro-ethics and macro-ethics. Is there a necessary tension between these two perspectives, and does this represent one of the tragic choice situations involved in living and surviving? Is there really a solution, or are there limits to rational analysis?

We Can Never Do Only One Thing: The Ecological Model

Case Study 5. Caudal Regression Syndrome in Son of Diabetic Mother

The mother suffers insulin-dependent diabetes mellitus and has three other children, one of whom has juvenile onset diabetes. The fifteen-month old boy has very small lower extremities, hypoplastic pelvis, vertebrae missing, limbs bent under, and club foot. There is an association between the mother being diabetic and this syndrome. Conservative incidence figures for offspring of

---

[16]See Richard Taylor, *Good and Evil*, Macmillan, New York, 1970, for an example of a philosophy that pays attention to other than rational aspects of human behavior.

diabetic mothers are 1 in 1000, while the general population incidence is 1 in 60,000 births. In general, in diabetic mothers, 8.1% have babies with congenital malformations. Some of the more common malformations include anencephaly, spina bifida, sacral dysgenesis, transposition of major vessels. There is a possibility that insulin may be a teratogen. Since we are not accustomed to thinking in terms of an ecological model in which there are usually other major effects in addition to that with which we are primarily concerned, we often view such additional effects as a basic flaw in the interventionist technique, as evidence that something is wrong with medical intervention. We really do not want to see things in terms of an integrated system since there are then always costs (in terms of unwanted effects) to any move that is made, and even to the choice to make no move. We call such effects that we would rather not have occur, or rather not pay attention to, "side effects." But this means no more than that they are not central to our purpose or interest, not that they are any less important than our principal effect. We quite naturally prefer to avoid more involvement than seems absolutely necessary in the ecological or systems model, and confine our analyses and operations to the more comfortable (because more simplistic) linear "one cause/one effect" model. Our popular and cultural perceptions exhibit a time lag in terms of new scientific models.[17] It would be more comfortable to have insulin treatment result in only one major consequence, the maintenance of a more normal metabolism, as indicated by blood sugar level, but it would also be highly unrealistic. The resulting reaction against the medication and the medical intervention is actually a reaction against our own denial of the realities of the situation. Such a reaction is still denial, however, because it again fails to face the multiple effects of not intervening and of not using the medication. To properly evaluate this case, a better mindset is required, one that recognizes another hard reality: that in a functioning system, there are very few free lunches.

Case Study 6. Family History
of Minor Hearing Impairment

In this case, a woman is pregnant, and has decided to have amniocentesis even before coming for the standard preliminary interview. During counseling, her concerns about a family history of

[17]H. Soodak and A. Iberall, "Homeokinetics: a Physical Science for Complex Systems," *Science* **201**, 579 (1978).

minor hearing impairment surfaced. She and her sister both have slight impairment. At the birth of her own first child, she had asked that the baby's hearing be checked, but this was not done, nor did the subject of genetic counseling for her apparently come up at that time. She continued to be worried, but never broached the subject until this present counseling opportunity, which was ostensibly for risks for fetal chromosomal abnormalities related to maternal age. This is a case of non-intervention at an earlier time, of "doing nothing" in terms of the mother's concerns and the family history. Did we actually succeed in doing nothing? Were there no consequences from this behavior? There appear to be some significant effects in terms of continued anxiety for the woman and lost opportunities to improve hearing had there been slight impairment in the children. The continuing effects of moderately high levels of anxiety or of functioning with a minor hearing loss could be carefully traced out and compared to the effects of counseling intervention and testing. The important point is that in an ongoing process, "doing nothing" has important consequences.

Case Study 7. Abdominal Mass Detected in Fetus

An ultrasound scan was done on a pregnant woman because of the uncertainty of date of conception. The scan showed a large abdominal cyst. Ultrasound scanning was then repeated, and a cyst filling the entire fetal abdomen was revealed. The couple is highly motivated to continue this pregnancy since they very much want a child. An alpha fetoprotein test was normal, which made the possibility of Meckel's Syndrome with polycystic kidneys less likely. The mass may well be inhibiting fetal growth, and may also eventually interfere with delivery. How much experimental intervention would be justified in this case? Should any intervention be considered? The obstetricians were considering an *in utero* tap of the cyst, but changed their minds. Cells from an amniotic tap are growing very poorly. The pregnancy is only twenty weeks along, and will probably result in a spontaneous abortion. At the moment, the doctors are waiting. There could be some possible risk to the mother. With such a highly-motivated couple, it would be relatively easy to get agreement for highly experimental and intrusive interventions. We have rather carefully to weigh the likely beneficial consequences of intervention and non-intervention in this situation. Does the role of the expert here call for some directive decisions about action or inaction to supercede the parents' decision?

Is non-interference in a process that will likely end in termination of the pregnancy the best we can currently do for this couple? How do we decide for inaction?

One of the hardest tasks in any search for knowledge is the identification of all the relevant consequences (seeing all the important implications). The ecological model has certainly given us an empirical feel for this task. Tracing out significant strands in the web of life, discovering the relevant interconnections, and looking at the major feedback loops in a system are important conceptual tools to apply to the study of any dynamic system. When medical ethics discussions involve an analysis of the balance between the good and undesirable results of various options or stances, even in so broad a sense as increasing the good in the world, the ability to delineate the unexpected effects, to trace out the indirect costs, and to see the long-range results is crucial. In ethics, in the end, we may all be disguised consequentialists. Rule and deontic positions in ethics imply there are some goods, usually termed "rules" or "principles," or "the right," that override all other consequences, that the result of violating them creates more evil in the universe than can be balanced by the good of violating them. Whether they are called duties, laws, or rules, the consequence of acting on them has a value that implicitly overrides all consequences of not acting on them, although this is usually phrased as a rejection of consideration of the consequences in ethical decision-making. This calculus may remain implicit because of tradition or obedience to a supreme power or to culturally accepted norms (and only surface, for example, when questions of something being good because an omnipotent all-good being commands it, or good because such a being must command it, are raised). However, at some past point, an explicit choice must have been made concerning the overriding worth of these principles *vis-à-vis* alternatives, and a decision was made now embodied in the tradition or in an assumption about intuition that the amount of good in the universe was greater in choosing these principles (or choosing to obey these principles) than in rejecting them. The valuation (choice), because implicit, is not as well-developed as most consequentialist positions and is frequently characterized as "consequences be damned." But of course they are not. The consequences of acting on such principles are clearly valued as greater goods than not acting on them; it is just that these consequences are rarely specifically articulated, and the results of the calculation are already assumed. The deontological position in ethics that I have been describing (that what is "right" is not determined by an examination of consequences) is a cost/benefit analysis that has been decided. So for *any* position, the identification of the consequences of various options is crucial.

An objection may be raised that Kantian ethics fits badly into my generalized consequentialism. If it does, it is because of value choices we prematurely assume in a Kantian system, resulting in the "good will" being necessarily vague. Noncontradiction assumes an overriding value, although the calculus of determining which behavior, rational, nonrational, or irrational, produces the greatest good has certainly not been worked out. Good in the world is equated with certain Kantianly necessary categories and operating in terms of those categories is to be valued more than empirical feedback of pleasure and pain, success in goal-reaching, or adaptation to external environment. The consequentialist choice is there implicitly.

Using the ecological model is really a pragmatic requirement, then, since we need accurately to identify the relevant consequences. It is a more sensitive model than the linear cause–effect model of the old popularized physics. The greater sensitivity is a result of the model's ability to follow out longer causal interconnections (to deal with "indirect" consequences) and to point out unexpected consequences. The Meadows' [17a] world system computer analysis, described earlier, would be a good concrete example of these attributes. Such a model seems better able to deal with the complexities of Open Systems, the category into which most medical situations fall. Our discussion of the importance of indirect costs and benefits in a cost/benefit analysis reaches the same conclusion.

Some rather famous examples of the surprising results of use of the ecological model are at hand. In general, Garrett Hardin has capsulated two of the basic operating rules: (1) We can never do one thing, and (2) We can never do nothing.[18] Because of the interconnections, some rather unexpected things happen when we intervene to accomplish one purpose. The more complete effects of preventing and fighting forest fires is a recent example from forestry studies. It turns out that the burning away of at least some undergrowth is necessary for a healthy, changing forest system, and controlled burning is now a new Forest Service tool. My dis-

[17a]Donella H. Meadows, Dennis L. Meadows, Jørgen Randers, and William W. Behrens III, *The Limits to Growth* New American Library, N.Y., 1972.

[18]Garrett Hardin, *Exploring New Ethics for Survival*, Viking, New York, 1972.

cussion of the Florida mangrove studies,[19] and recent analysis demonstrating the utility of some mangrove destruction in the ocean/shore interface is another. Certainly the rippling and unforeseen consequences of use of broad-spectrum insecticides is a famous one. The arguments here are not meant as a brief against intervention, but rather as a request for a more accurate characterization of the nature and consequences of intervention. That point should be made quite clear, and leads us to the second (and equally important) rule that non-intervention is an option that also has far-reaching and unpleasantly surprising results. Inaction is a behavior, a choice from which consequences also flow. In any ongoing system, processes will continue to occur whose consequences we must consider. Not introducing limiting factors is a major decision, certainly as major as introducing them. Our old static model had lulled us into thinking there was an unchanging status quo and that we had no need to consider the effects of inaction or non-intervention. It was as if not making an involuntary commitment, not performing an abortion, not doing research, not using an aversive conditioner[20] had no relevant consequences, but froze whatever was the current state as if it had no ongoing events. We should now know that inaction is better described as "continuing to act in the same way, rather than differently."

Attitudes Toward Pain and Suffering

The final consideration is that of what our basic attitudinal approach to pain and suffering ought to be. There are two different broad-scaled attitudes or valuations of the role of pain and suffering now underlying the various opposing conceptions of proper medical ethics. In a very general sense, these may serve to distinguish the religious and humanist position from the empirical and medical position. The first approach to pain and suffering may be characterized by acceptance of pain and the search for, or testament that one has found, significant personal meaning in pain and suffering. Pain becomes a positive value, an instrumental value necessary for the full development of the positive aspects of hu-

---

[19]Colleen Clements, "Stasis: the Unnatural Value," *Ethics* **86**, 136 (1976).

[20]For an excellent example, see Israel Goldiamond, "Toward a Constructional Approach to Social Problems," *Annual Review of Behavior Therapy*, Franks and Wilson, eds., Mazel, New York, 1975, pp. 54–57.

manness. There are some qualities of resignation and willingness to cooperate (at least by omission) in human sacrifice that are really two-edged swords, however. Of course, hopeless situations, problems that cannot be solved through change and manipulation, can only rely on this internalizing of the difficulty, that much is plain. The construction of meaning, in a perhaps meaningless universe, is a very natural human activity. It is often astounding what meaning can be stubbornly and grandly wrenched from absurd external circumstances. The humanist is rightly impressed with this very human behavior. It is one of our most important modes of self-definition. But it is not the only one, or the most obviously important one, and the consequences of attending only to this behavior can be frightening: a basic retreat from existence. Consider some of the possibilities. How does one define and determine a hopeless situation? How many situations become effectively hopeless because we perceive them as hopeless? Possibilities for the future can be effectively cut off by this attitude. Since even in systems theory, "futurism" is not the forecasting of an inevitable scenario, we may prematurely and arbitrarily close out all sorts of possible futures. The folk wisdom that "he didn't know it couldn't be done so he went ahead and did it" is not unrealistic.

Acceptance can also be problematic. The alternative value, that some things shouldn't be accepted, contributes as much to the meaning of human existence—to the same dignity, worth, and importance—as does acceptance. I doubt philosophy can adjudicate between them, the Stoics to the contrary. These are basic emotional responses, and determining their appropriateness to the actual situation is sometimes simple, but usually extremely complicated and prone to reduce to begging the very question at issue. However, even in simple terms of energy expense, acceptance often fails to supercede alternatives, since the emotional energy expended in maintaining internal acceptance of a painful circumstance may be considerable, though not always as obviously apparent.

Finally, it is necessary to be very careful about making pain a positive value; very callous consequences can readily follow. There are other behaviors that can elicit the full expression of our humanness, and too strong an emphasis on the value of pain can easily dilute our compassionate response to distress. Especially problematic is the instrumental use of another person's pain for our own individual or common good (in the sense of developing our humanness). For example, it is often argued that a severely

handicapped child had had the effect of bringing a family closer together, of making siblings more sensitive and compassionate, of
causing the spiritual or human growth of family members. Although such consequences sound desirable, what is really going
on is that the child is being used as an object, an instrument for
someone else's good. The child's pain is considered as a means to
the family's benefit. If that is the only consideration, the child is a
sacrifice. A positive value for pain has an instrumental rather than
intrinsic character, since pain alone is usually viewed as a natural
evil. This creates some difficulties in reconciling a natural evil and
an instrumental good, sometimes a delicate compromise or a conceptual stew. An instrumental good must somehow lead (and usually be the only alternative that will lead, if it is an intrinsic evil) to
an intrinsic good that supercedes the evil of the instrument. This
is, of course, the old ends vs means problem. A number of things
can go wrong: We can be mistaken; the intrinsic good may be less
than the evil of the means. There may be other alternatives that
lead to the same intrinsic good, without the use of an evil means.
We may confuse the instrumental good with an intrinsic good. I
think it can be safely said that the first approach to pain and suffering has fallen into all these errors.

   The second approach to pain and suffering is certainly a forth-
right interventionist approach, as the medical approach is aptly
characterized. It shares the optimism of science in general that our
perceived reality can be externally altered. Its major pitfall goes
back to our discussion of the ecological model: consequences are
usually far-ranging. But it is certainly another of the human species' basic behavior patterns: man the maker, the tinkerer, the curi-
ous, the manipulator, the dreamer, Faustian certainly, and not so
different in that from the rest of the primates.

   Our concern and care for each other, our social condition, mo-
tivates us to try to prevent suffering that we perceive as preventable. Standing by while aware that a specific pain or unhappiness
may be unnecessary is difficult to do. Medicine is not geared to
passive observation and the allowance of self-destructive behavior. This compassionate interventionism is a key feature of the
medical model, and is not to be lightly discounted. It is one of the
reasons a rights model will work so poorly in medical ethics and
one of the reasons that critics use the term "paternalism" in the pe-
jorative sense. It is also really not so simple as opting for patient
autonomy, or for the legal adversary role, as an easy answer to all
problems in medical ethics. The fundamental problem is actually
one of carefully ascertaining when efforts to help another avoid

pain and suffering may in the future rebound to cause much greater pain and suffering. To solve this problem by caricaturing efforts to help as heavy-handed paternalism or a plot to enable power-grabbing white-coated shamans to rule society is not a viable solution. There is, of course, clearly a need for patient participation in the physician/patient relationship. It is, after all, an interaction. The patient often needs to make the final choice (unless he or she is not functioning up to some minimum norm, the absence of which renders the whole notion of human choice meaningless). It is also an interaction, however, involving some agreement on the role of an expert in such social situations. Defining away the function of the expert in a social system does not seem to me a real alternative and yet is apparently implied in some of the ethical literature calling for change in the physician/patient relationship. I will look at this in more detail later. For now, the significant point is that characterizing the problem in ecological terms (are the consequences of attempts to prevent pain and suffering a greater evil than the consequences allowing the pain and suffering) more clearly indicates what the difficulty actually is. With such an understanding, inappropriate solutions become easier to point out. Such solutions usually contain, as their unexpressed bottom line, a rejection of this second basic approach. Since it *is* unexpressed, the consequences of such a rejection are not considered. Again, ecologically, or even existentially, we can never do nothing, we can never choose nothing.

Although in the medical model, pain can be a useful marker, a diagnostic instrument, it is not something to be accepted, borne, or resigned to. Even in terminal illness (as for example, in the hospice model), an attempt is made to eliminate pain as completely as possible. A state of continuing pain is unacceptable. As a result, the medical approach is a committed, constant attempt to prevent pain and suffering. One of the problems is then obvious. When does one reasonably accept defeat, give up, in treating pain? Although the first approach may prematurely allow acceptance of what seems inevitable, this second approach may not accurately judge when the struggle is really useless. If there were no major costs in such a quixotic struggle, we would surely prefer to err on the side of trying. But there are always costs, and sometimes the costs become very high. Again, using terminal illness as an example, the costs to a patient when the health care system refuses to accept that patient's dying are considerable (the pain and discomfort of heroic interventions, the psychological stress, the ignored consequences to the family system). I believe we can safely say that there are costs for either basic attitude and that both behaviors are certainly key human possibilities. But I am still tempted, in terms of a preference for open systems and negative entropy (greater complexity and activity), to think that the second approach represents a dynamism and future potential that has much to recommend it. Since biological organisms are open systems, and since static systems tend to disintegrate more rapidly in interaction with other systems, I feel more comfortable with the second approach. Although it entails appreciable risks, we are not locked into a static pattern of response. It is an activist approach that is compatible with the adaptive change that is a feature of the history of biological organisms.

The points I wish to emphasize here, however, are that both approaches exemplify typical human responses to pain (or any other event), and especially that the interventionist behavior characterizing the sciences is an expression of our humanness (humanism), having all the dignity and value of a significant human response. Both approaches have costs and limitations; both are expressive of equally important aspects of our human nature or human possibility. Medical interventionism is also humanism.

Chapter 2

Working with Information

There is an implicit human valuing of information, one embodying religious formulations that the truth will make one free, philosophic formulations that the unexamined life is not worth living, scientific premises that knowledge is power one can use for human good, psychoanalytic principles that ego (usually viewed as logos) should replace id, and common sense principles that one should look before leaping. I would like to take a closer look at how we make this choice for information, some of the justification for such a choice, and the implications of this choice or valuation. In medical ethics, it is not always as clear-cut as the above stated axioms make it seem.

Specifically, in a prenatal detection program, I want to approach this problem in terms of a negative situation: working with information gaps. There are four general types of insufficient information that frequently arise in the clinical setting.

(1) Lack of documentation, verification of diagnoses, or autopsy to establish diagnosis.

(2) The possibility of one syndrome or of separate and unrelated problems.

(3) The use of catch-all explanations.

(4) The appearance of new and unreported syndromes.

How the search for information, the use of available information, and the lack of information affects the physician/patient interaction in each of these situations may help us more clearly unravel the ethical character of information in its more general sense; what is really going on, in other words, between our rational, nonrational, and irrational behaviors.

Lack of Documentation, Verified
Diagnosis, or Autopsy

Case Study 8. Possible Translocation Trisomy 21

The patient is 16 weeks pregnant and concerned about a possible family history of Down's syndrome. In terms of age, she is well below the maternal age risk for Down's syndrome and has two normal children. Her mother had four early spontaneous abortions and the patient has only one sibling, a brother who has, according to her, Down's syndrome. Her mother's sister, at 45, had a son with Down's syndrome. Medical records on either Down's individual were not available. The patient was referred by her doctor, who was very directive in this matter. There are some problems with information in this case, but the required information can be obtained. The concern, of course, is with the possibility of translocation rather than maternal age-related Down's, and the maternal aunt's child is probably not relevant. The mother's early abortions and the Down's brother would be the indicating factors. This brother, now adult, was never karyotyped to confirm the suspected diagnosis. Arrangements were made to do that, as well as to karyotype the patient. Unless the typing indicated a balanced translocation, with the patient a carrier as well, the genetic counselor did not feel amniocentesis was indicated, except possibly to reduce anxiety. This is a relatively straightforward case, since the needed information can be relatively easily obtained. The medical intervention necessary to obtain this data is not intrusive. A simple blood sampling is required; the blood cells are then cultured, and with banding techniques, a quite reliable karyotyping can readily be accomplished. The mother would have to serve as surrogate to give consent for the brother's testing. After the test, the patient could be reassured either that she was not a carrier, or that if she were, amniocentesis could then be done to determine the status of the fetus. Taking the blood sample from the brother would not benefit him in terms of treatment, of course, but we presume he has normal emotional relationships to the family and would probably wish to help his sister. Certainly the mother would have caring concerns for her grandchildren, unless the family structure were quite pathological. Information that the patient was a carrier for a trisomy 21 translocation could alter her perception of herself, of course. There is a 10% risk to the patient, if she were a carrier, of having a Down's child. Is this risk greater to her than a change in how she perceives herself? Information has many effects. The patient now has been counseled that a balanced translocation is a possibility, and this is also information. Earlier the patient had only vague anxiety about her family history (anxiety reinforced by her referring physician), but she now has a very specific anxiety. Being karyotyped can relieve or confirm this anxiety. Refusing karotyping and further information will maintain the anxiety. In all the many ways that information can be conveyed, it is certainly a significant intervention. However, the decision not to give information also has significant conseqences. In this case, not giving information could maintain the anxiety produced by the family's experience with Down's syndrome and could involve accepting at most a 10% risk of partial responsibility for the patient's coping problems with a Down's child.

Case Study 9. Possible Turner's Syndrome

The patient is a 12-year-old girl, small for her age. She is here for genetic counseling to check the possibility of Turner's syndrome, a sex chromosome abnormality (XO) associated with small stature, sterility, webbed neck, and sometimes mild mental retardation. There is no endocrinopathy. Her mother is also short-statured, 4'11". A sex chromatin test, initiated by her pediatrician, indicated a possible sex identity problem, but such a test is frequently unreliable. The patient is here for chromosome analysis, basically to clear the air, since the genetic counselor suspects the sex chromatin test was in error, and that the diagnosis should actually be familial short stature. The girl was repeatedly told she was normal and the parents were also assured. Enough information has already been given (the pediatrician's evidence of concern in requesting a test, the sex identity question resulting from the test) to produce significant consequences in this family's life. The child's whole sexual image is in jeopardy, at a very critical time in her life, as a result. In fact, the medical management of this case up to the point of counseling is probably iatrogenic, and the counselor's task may be to try and correct a medically induced problem. Should the pediatrician, in the absence of any endocrinopathy and at too early an age for confirmed primary amenorrhea, have requested a sex chromatin test? Should a pediatrician have known enough not to request such a test? Would a general practitioner not have known enough to request the test (even if there were sufficient indication for it)? And which is better medicine: to be trained well-enough to know when such tests are indicated; or to be insuffficiently trained to know the availability of such tests and have to rely on a feel for

discriminating problems based on accumulated but limited experience? Or is either good medicine, and need we be limited to such a
choice?

In this case, premature imparting of probably irrelevant information could have a serious iatrogenic effect. The information
could be a disvalue rather than a value. Additional information is
now required to relieve fear and prevent significant damage to the
child's sense of self-worth. The ability to be comfortable with her
sexuality and to have a pleasing sense of body-image could have
been compromised. Her parents' relationship to her has been put
under stress. The information from the chromosome analysis may
relieve this, as may a reassuring interaction with the genetic counselor. In this case, we are fortunate that a reasonably definitive answer can be given, although the possibility of Turner's Mosaicism
can *not* be completely ruled out. Should this possibility even be
mentioned to the parents or the child? If it is mentioned, in light of
the counterindication for doing any of the tests in the first place,
would it destroy the value of the information obtained from chromosome analysis? Can information have a significant enough disvalue to warrant choosing against it, and who should do the
choosing?

Case Study 10. Karyotyping
to Confirm Down's Syndrome

A pediatrician referred a 2½-week-old baby girl to confirm suspected trisomy 21. The baby had a protruding tongue, epicanthal
fold, and functional heart murmur. In some cases, confirmation of
diagnosis could be necessary for state aid. Federal and state regulations may increase the necessity for gathering information, and
we then have to trust that such regulations take accurate account
of the value of information. Studies also indicate that there is a better adjustment to Down's syndrome when karyotyping is done,
that this somehow objectifies the problem, separating it from the
parents and giving them a sense of being not at fault in terms of
their total selves. Confirming a diagnosis also gives an answer or a
label to the painful experience. Knowing, in the sense of naming
and understanding, seems in itself to have a therapeutic effect.
This may arise from our fear of the unknown and unfamiliar, of
blind fate or chance. On the other hand, knowledge can give us
the internal power to cope, as well as the external power to
change. Finally, confirming this diagnosis could be a possible future aid to relatives. Since this test is not intrusive and merely confirms a problem rather than first pointing one out, it is not prob-

lematic. How significant would the benefits have to be, and to whom, if it were intrusive?

Case Study 11. Central Nervous System Abnormality

The patient is 4 months pregnant. Her sister died one day after birth with a "central nervous system abnormality." There are "slow" children in the family history, with no documentation on school performance. The patient came chiefly because her obstetrician was concerned with the family history. She explained that her mother was "neurotic," considered family pedigree private and did not want the medical records released. She also asked if the test would help her fetus, in the sense of preventing a genetic defect, and was told it would not. She did not appear to value information or feel it could give her reassurance. However, she did put pressure on her mother to release her sister's medical records, and the mother did sign a release. At the same time, the patient refused a blood sample chromosome test for herself. The resident physician questioned the nondirective approach of the genetic counselor as opposed to what he perceived as the directiveness of physicians in other specialties where he felt unilateral decisions were commonly made. The geneticist felt two steps should be taken: (1) more information had to be gathered, and (2) an evaluation had to be made of how the patient would handle risk information.

There are ambivalent signals coming from the patient. She has pressed for information about her sister, but she is resisting information about herself or her fetus. Information seems both to have, and not to have, value for her, and her choices are inconsistent. The patient's mother also seems to have a problem assigning value to information. She preferred to keep the sister's medical records private. Does she have the right to keep this information suppressed, had she chosen to do so? The values of information and privacy often conflict. How do we choose between them? Was the patient's pressure on her mother justified, in view of her own denial behavior concerning chromosome analysis? How much denial should a physician accomodate?

Case Study 12. Huntington's Chorea

The patient is a self-referred 24-year-old woman. Her father was hospitalized at 47 with probable HC, and her grandfather also possibly had HC. She had not been contacted by any medical source about this and knew a small amount about the genetics of the disease. She had had to take the initiative to obtain any information or genetic counseling. Her father's diagnosis has not been documented and definite information cannot be provided to this patient on the basis of incomplete data. There are two questions in addition that need an answer: (1) What precipitated this present contact? and, (2) Why is she receiving no therapeutic support? If her father's HC could be confirmed, what information can we give her? Basically, that she would be at 50% risk herself for HC. Could we lower the uncertainty figure? L-Dopa is not being actively pursued as a diagnostic test. If a diagnostic test could be developed, how would people value this information? A survey by an HC group indicated that 80% would want such a test, while 20% would not. Again, knowing, even in the face of the reality of no treatment and the progress of the disease, seems to be valued over the anxiety produced by not knowing. That this occurs in a situation where one is at equal risk for having or not having the feared disease is an important statement of the value of information. This is not, however, an indication of how a person will deal with the information, only a statement that he or she values the information. The patient may deal very drastically with it, to the point that some HC-indicated patients have committed suicide. If a test became available, the 20% who did not want such a test might further reduce to a certain percent who would have the test done because of the pressure of its availability and the remaining percentage who would still not value the information. The benefit of knowing one does not have HC is apparent. What are the benefits of knowing one does have it?

Case Study 13. *Mental Retardation and Dwarfism*

The patient is a 34-year-old with severe mental retardation, epilepsy, and dwarfism. He is a twin, with the other twin normal. The patient requires total patient care and does not communicate. The diagnosis could be possible endocrine abnormalities: cretinism and hypopituitarism, for example. There is no indication from the referring institution of the reason for referral, and management of this case would depend on the referral motivation. Do we need a definitive diagnosis in order to give the normal twin information he would like in order to make reproductive decisions? Does the institution need a diagnosis to meet some regulatory requirement? How intrusive a work-up should be done? Would the information gained be of any value to the patient? Does the family need this information and how great is the need? If this retarded patient has "rights" under ethical models based on legal analogies, does he also have "responsibilities"?

Case Study 14. *Megalencephaly and Hydrocephaly*

This couple was referred after the birth and subsequent death of a baby finally diagnosed and confirmed at autopsy as having megalencephaly rather than hydrocephaly. A nephew has been diagnosed as having hydrocephaly, but it is difficult to rule out megalencephaly. The baby's skull was oversized and the brain was almost the size of an adult brain. The baby had psychomotor retardation and died at 10 months during a pneumoencephalogram (an air embolism developed, causing acute death). A CAT scan, of course, would not have carried this risk, and a CAT scan has been suggested to confirm or disconfirm the nephew's diagnosis. It is important to determine, for purposes of counseling, whether the two children both are megalencephalic. If that were the case, a dominant mechanism would be more likely and the couple's risk of affected offspring could be as high as 50% in each future pregnancy. They returned for follow-up counseling. The wife has one child by a previous marriage. With still no definitive diagnosis on the nephew, we cannot provide them the information they need and want.

In this case, autopsy finally established one diagnosis rather than another in the propositus, indicating the importance of this medical procedure, especially in terms of future decision-making. In the case of spontaneous abortions, autopsy information is also genetically very useful. Should hospitals routinely run certain tests on tissue from early abortions? In this case, were the nephew to die, what obligation would his parents be under to see that a detailed autopsy were done, or would they have none? There may be two motivations for not performing autopsies routinely: (1) the cost of the procedure if instituted on a wide scale, and (2) feelings that "identity" is still invested in the individual's body after death, and that highly intrusive and mutilating procedures are therefore still assaultive to that individual. A corollary to this would be the psychological stress that family or close companions might experience in terms of how the body is handled. These may frequently outbalance the value placed on information, even when that value is fairly high. Should they, ethically or conceptually?

Any gathering or imparting of information, we have seen, is an active and interventionist process, so that by definition the values of information and privacy conflict. In addition, it is usually impossible, also by definition, to give patients an informed choice on whether they wish information to be conveyed to them—the informing already has done the conveying. There are two major conflicts, then, built into the information process—conflicts concerned with privacy and autonomy. The real territory of ethical analysis is at the interface of those (or other) conflicting values, if it is to be a particularly useful analysis. These conflicts determine just what our valuing or choosing can really mean. Although it is easy to say that one always values "truth," these conflicts begin to define just what one is willing to accept as "truth" and what one is willing to trade away.

Privacy implies that there are matters we do not choose to have others know, or to go about finding out. Since information is one of our society's main products, it becomes harder and harder to restrict the accumulation of information, so we fall back to a second line of defense, attempting to control the dissemination and use of that information. In doing this, we can depend on both legal and fiduciary tradition, although I tend to feel that the fiduciary ethic on confidentiality is a better safeguard than the legal sanctions supporting it, simply because (1) this ultimately is a human interaction and the law cannot foresee and regulate all possible behavior, and because (2) of the real limits of the law's aversive conditioning.

The first question at issue is what value privacy has for us—why we choose it. The second is to determine, when privacy and information values conflict, what steps we generally take to define these two values to modify the conflict, and in a last resort, which we choose when compromise is no longer possible. However, what is of philosophic rather than (or better, in addition to) anthropological interest is that there are rankings that are individually or culturally relative, and rankings that can be hypothetically objective. In addition, it is sometimes claimed that there are rankings that are categorically objective. How would this ranking hierarchy operate when we attempt to rank information relative to privacy and autonomy? We could certainly do a study to see how people in value conflict situations perceive the conflict, what behaviorally they mean by these values, and how they go about ranking them. We might or might not find a ranking relativism; the important thing at this level of analysis is that we *can* carry out such a study even if we do not. The next step in ethical analysis is to see whether the "can" may be eliminated, and then to determine which functional norms—certain natural (in the sense of always occurring) characteristics—to use as prefaces to a ranking choice. For example, the study might include questions such as:

> If you wish to remove anxiety, then _____.
> If you wish to be healthy, then _____.
> If you wish to avoid pain, then _____.

These "if clauses" usually can be generalized to concepts of well-functioning, survival, and species survival. There has been some attempt to create a naturalistic ethic on this basis, to assert that these generalizations should be and/or always are the valuations made. The Aristotelian and natural law formulations, the hedonic perspectives, are in this tradition. They attempt to move to a categorical ranking, but such a move presupposes an invariant environment, an invariant human nature, and an invariant relationship between the two that equals well-functioning. Such a presupposition is unworkable. It violates what we know about processes or systems, what we know about reality-as-we-perceive-it. Bringing in an unknown reality to rescue the presupposition is a useless move.

Nevertheless, this step of determining norms has validity. It's part of the answer, but the level of analysis needs to go to a more basic ground. Values such as well-functioning or survival make sense in the context of an affirmative affect (response) to experience, but not in the context of a negative affect. Prior to valuation (ethical, cognitive, esthetic), a person has an affect or attitude concerning his or her existence in the world, an affirmative or negative attitude, and it is the affirmative attitude that is the precondition for all valuing. Such affirmation or negation is a primary affect that cannot be justified or rejected philosophically since philosophizing itself necessitates the acceptance of cognitive values that are founded on the affirmative attitude.

Who cares about actual and possible pleasure or pain, about surviving and functional potential, about the future of the human species? Those who affirm existence—the quick, and not those who wish to be dead. This added component, the interaction between a changing environment and a potentially changing response to that environment, makes a categorical ranking impossible. But it accurately expresses the homeostasis involved in our human natures and the evolutionary character of the physical world. Ultimately, then, the ranking is hypothetical: *If one affirms existence (experience), then that person will try to maintain functional norms (in new or old experimental ways), and these are the choices, to the best of our knowledge, that will effect that goal or purpose.* That latter clause can be reasonably empirically determined, so that the adoption of such an ethical position is not relativistic. The "if clause" gives us the option of reacting in a systems-maintaining or systems-disintegrating way, both of which activities characterize the universe as we know it, and neither of which seems to have any "natural" ranking above the other.

In our specific case, we can in a sense objectively rank privacy, autonomy, and information. Given the affirmative stance and the value of well-functioning it implies, we will in most situations probably decide that reality-testing (information) is superior to denial. Although information raises short-term anxiety, it is an effective tool in preventing long-term anxiety and allowing us to devise more ways to maintain stasis (functioning and survival). It has some significant costs that cannot be ignored, however. Privacy allows us to maintain individual, unique functioning that may not conform to social norms. In some cases, this functioning is not only socially deviant, but self-destructive. But in many other cases, it is the source of creative problem-solving and innovation to meet changing environmental pressures. These also are necessary to maintain stasis. In most medical situations, however, privacy does not serve this important function, or when it does, the role's tradition of confidentiality serves as sufficient protection for this function to continue. The privacy of the Down's sibling is not under serious threat by asking him or his parent to allow determination of the differential diagnosis of balanced translocation. Even a routine state-prescribed screening really does not affect our ability to function in socially varying ways. Family pedigree information (e.g., the extensive records of the Mormon Church) does not impose a conformity in terms of reproduction. Learned attitudes and the social uses of the information can be a different story, however; it seems clear to me that this is the only practical area of restriction, and does not justify a censorship of the information itself (a denial choice).

The cases of Turner's syndrome and HC reflect problems of self-attitude, of self-perception, that are often learned from inaccurate social models. A re-examination of these models, and support for those afflicted with these diseases until perceptions can be altered, is more feasible than information restriction. An individual counselor, however, may have to balance current value and disvalue when privacy and information conflict, and when the effects are serious and presently irremedial, may have to choose to be non-informing (always, one hopes, only as a temporary measure).

Privacy as a valuation can be defined, then, in terms of the possibilities it affords for variant functioning. It does not involve a nonsocializing characterization, however, nor allow us to view humans as nonrelated, unknown, and remote individuals. Privacy is not isolation. Confidentiality, for example, is actually understood in a context that weighs social relationships with the patient's need for variant behavior. Thus confidentiality is breeched when it is judged that the need served by the revealed information could be met in ways not so seriously destructive to others in the individual's net of relationships. It is uncomfortable to make these everyday, working choices. It seems easier to have a blanket proscription, an inviolable rule. In day-to-day living, we're smart enough, nevertheless, to know the trail of anguish such unresponsive and unrealistic rules create. The legal system is an excellent example of the constant modification and evolving interpretation of any law.

Information also affects autonomy, but in an extremely ambivalent way. A choice made in ignorance is not a choice among actual alternatives, and certain philosophic perspectives, e.g.., Existentialism, would define it as no choice at all. Autonomy, in that perspective, implies understanding. Notions of informed consent clearly illustrate the connection between information and autonomy. Yet information has other effects that also are related to autonomy. There is the short-term impact of raising anxiety. People's ability to handle such anxiety can vary considerably. Some cope so poorly that the anxiety becomes a counter to autonomy, freezing action, creating self-destructive behavior, precluding rational or identity-consistent choices. Instead of setting us free, the truth can sometimes lock us into nonfunctionality. If information has value in the sense of contributing to well-functioning and survival, it loses that value when its consequences are the opposite. This is what makes HC such a difficult problem. For many, the information that a definitive early diagnosis could give would allow them to make choices contributing to living fully for the period of time they could. For others, such information would make existence untenable, would create unmanageable anxiety, would reduce them to a nonfunctioning or malfunctioning condition. The Turner's case illustrates the same difficulty, in addition to the possibilty of misinformation and unnecessary intervention.

Autonomy has frequently been depicted as the "right to know," yet we can see that in certain instances autonomy is reduced or destroyed by knowing as well. At the same time, autonomy appears to make no sense when choices are made in ignorance. Again, the "rights model" will not get us very far here. We may have an interest in knowing, since knowledge seems integral to our well-functioning, and in that sense we have a need to know, when knowledge is actually integral to our well-functioning. We make the assumption that usually the anxiety produced by information is short-term and bearable, or that it can be modified through coping mechanisms and support systems. We do this because information usually has long-term effects that assist functioning and survival. It is, after all, one of our basic evolved means of interacting with our environment. Our evolutionary future has been staked on it. In the long-run, then, it usually contributes to autonomy, if we mean more by autonomy than capricious, wilful choice. Based on our hypothetically objective ethics, it is this broader concept of autonomy that makes sense. Restricting options through information, then, may violate the narrow notion of autonomy as purely free, with non-influenced choices; but then also, increasing options through information would violate that notion as well. Neither informational-imparting activity need conceptually conflict with the broader concept of autonomy. In practice, one is stuck (it seems to me irrevocably) with having to determine empirically whether in any individual case information destroys possibilities for an individual to adjust, or assists that process. The presumption is, however, that additional information will usually contribute to eventual well-functioning. It is important to determine, still, in which cases that will be an incorrect presumption; then, information will be a disvalue, not a value, and the price of informing will considerably overbalance the price of non-informing. We need continually to remember that both have a price in the real world.

One Syndrome or Separate Problems

Case Study 15. Upper Limb Deformity

The baby boy has no finger or thumb differentiation, but the general prognosis is good. The pregnancy was complicated by toxemia and a possible vaginal infection. Family history-taking showed that the father has Sprengel deformity, an upward displacement of the scapula, although this congenital problem had been presented as the result of an accident. The mother has considerable guilt feelings, and there was a question in her mind concerning the association of the two deformities. Sprengel deformity is not associated with the baby's deformity; however, it cannot be determined whether the child's problem is inherited or congenital. The mother showed a poor understanding of the physician's explanations, which were generally reassuring, that she has a very slight chance of this deformity recurring.

In this case, separating the two problems was relatively easy. It was also fortunate, since the mother had been given misinformation about her husband's condition. This case, in fact, supports the general value of information in terms of long-range consequences. It was, however, complicated by the mother's poor understanding of the information given to her. Imparting information is not a simple process; as in all interactions, both parties influence the outcome. The physician has some control over what is imparted, and in the role of expert, considerable responsibility for that, but it remains a two-way street. Information is not a precisely determined quantity poured into an empty vessel (or *tabula rasa*, if you prefer). Even when separating the two kinds of deformities is clear-cut in the physician's mind, it is sometimes very difficult (but also very important) to separate them in the patient's mind.

Case Study 16. Nystagmus

The patient is a 4-year-old boy with nystagmus. The family history is negative, with no indication that any male relatives have the problem, ruling out an X-linked disorder. His development has been normal, and an older son is normal. Since nystagmus is involved as part of many syndromes, the physician felt a complete neurological workup was necessary. The motivation for this was to bring the matter to some conclusion, to some decision on the parents' part, and to disengagement of the child from the medical system. Either positive or negative information could accomplish this and remove the boy from the "sick role" or place him in a labeled sick role. The status quo had him in a highly ambiguous position and could affect many of his social relationships. Since nystagmus alone is relatively innocuous, the information required is whether it is an isolated finding or part of a more serious syndrome. In this case, the parents are already sufficiently anxious that further intervention appears justified to clarify whether the condition is isolated or involved in a constellation. In other cases, raising the possibility of broader involvement may greatly increase anxiety, require increasing intervention, and not be productive. A wait-and-see attitude, on the other hand, may postpone discovery of life-threatening problems or allow reproductive choices the parents may later regret and resent. A categorical rule about information gathering or dissemination would often result in unnecessarily generated anxiety and stress for the family, needless and potentially dangerous interventions, and wasting of time and resources (in addition to being philosophically baseless as a value, as I pointed out earlier). A clinical or expert judgment, on the other hand, is open to human error, but also to human weighing of consequences.

Case Study 17. Connective Tissue Disorder

A couple was counseled after the post-delivery death of their first child. The infant had multiple fractures and cataracts, but no heart murmur. A connective tissue disorder may not have been the baby's major problem, but no autopsy was done and the diagnosis is therefore unknown. The wife has hyper-extensible joints and a family history for this. She has mild scoliosis and possible click murmur. Her mother takes quinidine for a heart murmur. There are a number of spontaneous abortions in the pedigree. The wife's brother had a son who died with severe extensibility. The origins of both sides of the pedigree go back to a small village in Europe. The simultaneous occurrence of hyper-extensibility and murmur may indicate that there is a new syndrome, and there is no previous information on this available.

The couple were asked to inquire very broadly about family problems, to attempt to gather as much pedigree information as possible. They appeared to indulge in at least some denial. This is a complicated case, because two problems may exist. Was the infant's fatal defect part of the pedigree syndrome, or is it a separate defect; and if the latter, what is it? If it is actually a separate problem, a recessive genetic mechanism or a mutuation might be involved and recurrence risks might therefore be low. If it is part of the hyper-extensibility syndrome, the mechanism might be a dominant one, with higher recurrence risks. A great deal depends on information, but the prospects for determining whether this is all part of one syndrome, based on present data, are not good. One chance to obtain important information, autopsy of the infant, has been lost. If the couple do try to add to the detail of the family history, they will be expanding the information-gathering intervention to the extended family. Does the family system imply such a shared responsibility, as a social ethic? How is access to information built into our social roles?

Case Study 18. *Aniridia*

The patient is a child over 4 years old with aniridia, the absence, usually in one eye but sometimes in both, of the iris. The physician can tell the parents with some degree of probability that the defect is not part of a broader syndrome associated with a deletion in chromosome 11. Such an association usually develops before 4 years of age, and the present patient has been brought in only past that age. However, chromosome analysis might have determined the disease sooner, and is important. In the broader syndrome, Wilm's tumor and nephroblastoma could develop. Wilm's tumor can be successfully treated, however, so in children at risk, monitoring should be done during the four-year risk period. It was not done in this case, nor was the symptom established to be independent of the Wilm's tumor/aniridia syndrome. Fortunately, this does not appear to be a case of the broader syndrome, but it might have been, and the lack of information could have had life-threatening consequences.

Given that in most cases information can be ranked very high in our valuation hierarchy, in my hypothetically objective sense, some of these cases also illustrate another problem: What concept of information are we working with? We have seen that information tends to be the product of an interactive process rather than a hard and fast entity. And there is certainly a considerable philosophic problem with what we know and how we can know it. A discussion of epistemology is out-of-order in a book on medical ethics, but I want to call your attention to it because it underlies various contemporary discussions of medical information, such as that on disease-labeling from Sedgwick to Szasz.[21] The criticism of disease-labeling tends to be that it is valuational, that our data about reality are subjectively-tinged, normative as opposed to objective, and therefore. . . What follows "therefore" is rather difficult to determine. In some cases, the normative is contrasted to a pure, perhaps foundationist, datum untouched by human hands. The old non-normative (nonvaluational, non-interest generated) *is*

[21]Peter Sedgwick, "What is Illness," in *Contemporary Issues in Bioethics*, Beauchamp and Walters, eds., Wadsworth, Belmont, Ca., 1978, pp. 114–19; Thomas Szasz, *The Myth of Mental Illness*, Harper & Row, New York, 1961; "Concepts of Health and Disease," *Journal of Medicine and Philosophy*, 1 Sept. 1976; Richard Hull, "On Getting 'Genetic' out of 'Genetic Disease'," *Contemporary Issues in Biomedical Ethics*, Davis, Hoffmaster, and Shorten, eds., Humana, Clifton, N.J., 1978, pp. 71–87; "The Concept of Health," *Hastings Center Studies*, 1, No. 3, 1973.

and the old, separate and non-empirical *ought* still underlie our formulations. Otherwise, what difference would it make that the concept of disease is value-laden, derived from norms, connected with the concept of well-functioning and stasis maintenance? Implied is that there is some other kind of knowledge, a better kind, that is value-less, does not use the concept of function, does not depend on our interests and purposes, or on our receptivity; as Piaget said, the object in our poorly formulated subject–object dichotomy version of epistemology.[22]

That we need to posit a reality over and above our human experience is granted, even demanded. The *Ding-an-sich* (thing-in-itself) has a crucial role to play in any scheme that hopes to be reasonably adequate. Human stipulation, human categories, persistently bump into unyielding limits (brute fact, hard reality), much as Johnson's toe would have bumped into the rock even if he had not been trying to make a point. Some human ways of experiencing work better than others (cause less pain, lead to wish fulfillment, meet our interests). This limiting (and Kant called the thing-in-itself a limit[23]) appears over and above our body of accumulated perceptions and habits of perception as a changing and often surprising restriction or expansion. Given all the varieties of human experience, the matrix for these intricate complexities is that limiting factor that is independent of our emotional and rational desires, but not of us, since as the pragmatists saw, we are part of that reality, *are* it ourselves. The relationship is truly a transaction.

If my excursion into epistemology is forgiveable, and if all the above is so, then the wrangle over objective and normative disease concepts is to repeat history unnecessarily in medical ethics. In our case, the imparting of information can be both a social and cultural process (normative or value-laden), and pragmatically justified relationship with limiting reality (objective, in the sense of not created by our wishes). It becomes a product, then, in a very different sense from a manufactured product. The patient makes a real contribution to the end result. Informing is working with.

Many of these cases illustrate another ethical and conceptual difficulty: determining what is necessary and unnecessary intervention. Should the patient remain in the medical system? Should we increase the patient's involvement in the medical system? How

[22]Jean Piaget, *Insights and Illusions of Philosophy*, Wolfe Mays, trans., New American Library, New York, 1971, pp. 39–77.

[23]Immanuel Kant, *Critique of Pure Reason*, Norman Kemp Smith, trans., St. Martin's Press, New York, 1965, pp. 270–272.

do we determine when intervention begins to cause more harm than good? Overtreatment is always a possibility, what Dr. William Greene calls "gang medicine" and "inflicting survival." Some types of intervention seem innocuous enough, e.g., calling attention to and checking out the 4-year-old's nystagmus. But in all interactions, something does transpire. Information is given, which has results, or information is found to be lacking, which usually involves going to the next step or considering giving reassuring, supportive help. Other types of intervention are highy intrusive physically, and medical experts can give them some reasonable ranking order in terms of their inherent risks. Should the patient, through informed consent, determine the level of intervention? Again, we have to keep in mind that we are discussing a relationship, and one member of that interaction does have expert status in terms of the physical risk of intrusive intervention. Not only that, but the expert member will have the responsibility of performing that intervention. If the risks considerably outweigh the benefits in the professional's mind, and not in the patient's, there is not only a conflict of interests and ranking of risks, but a conflict involving the very role of a professional as an expert. A good lawyer, for example, does not often collaborate in the legal folly of a client. Rather, the lawyer seeks to be excused from the case or attempts to get an incompetency judgment. A good physician has the same expert recourse, and is not required to collaborate actively in an interventionist folly; nor is such collaboration often wise. There are times when the physician may judge that some of the patient's other needs override the specific medical judgment that the physical risks outweigh the physical benefits of some procedure. For example, a surgeon performs the last of a series of operations on a badly compromised patient with little hope of success, knowing the patient is fully aware of the situation, wants to gamble in order to survive, and desperately does not want the only alternative of slowly and uncomfortably starving to death. Expert medical judgment must take account of the patient as a complete person, as I maintained in the first chapter.

Does the physician have passively to collaborate in the folly of a patient who refuses intervention? Incompetency procedures, of course, can be instituted and here the question of fully determining the consequences, of accurately assessing how much pain is prevented and how much pain is newly added, is vital. Will we *really* leave the patient better off if we force intervention, and what is the price of that force? It has a high price, in terms of autonomy, dignity and self-image, trust, and changed social relationships.

The impact is major. But the price for not forcing intervention may even be higher in some situations, and therefore, once again, a situation that may be difficult to interpret in a strictly legal model is encountered, and a "judgment call" is involved. In the literature of medical ethics, there is some reluctance to trust the medical community with this judgment call. The alternatives do not seem to me to be any brighter. We can continue to add to the legal role's activities in our society (legalization of our lives may be just as rampant as medicalization seems to be to some), or we can opt for a totally unrestrained relativism of choice (everyone can choose as he pleases). What we forget with the first alternative is that the legal community's method of problem-solving and decision-making is either strictly and traditionally codified, or else involves the same "judgment call" approach as does that in the expert medical community. And what we forget with the second alternative is the painful social and individual cost of allowing anyone to choose self-destructive behavior. Those considerations need to be heavily weighed in any proposed modification of the medical role, and usually are not even identified, let alone weighed.

Intervention that imparts information can be as problematic as physical intervention. The patient's way of refusing such information is through denial (not wanting to hear, not hearing, garbling the message, forgetting the message). In the face of strong resistance and refusal through denial, the physician has the same options as in the physical model, and I will discuss this in some detail in the next chapter. Complying with a patient's request for information that the physician feels is medically contraindicated, or acting on the notion that informed consent always requires total and detailed information conveyance, is analogous to actively collaborating in the patient's folly. The New York State Public Health Law (Sec. 2805-d, 4d), for example, has specifically codified this example of the disvalue of information and of the physician's role as expert (or at least, part of the interchange). The physician, under this law and for professional reasons, can decide not to inform. Again, autonomy through information is *not* always the consequence of conveying information. Rather, serious inability to function in necessary areas may be the result, and in that case, information loses its value-base. Truth for truth's sake is probably at best only a half-truth. We pay lip-service to such a valuing of Truth because in most cases reality testing does allow us to function at some minimum norm over the long-run. Most neurotic defenses, based on denial, do eventually break down or cause painful malfunctions. But "most" is not "all," and we need to keep in mind that information is an instrumental value, an evolutionary experiment in maintaining a complex system's stasis, and its ultimate justification is the pragmatic criterion. Treating informed consent as an intrinsic value sometimes places the physician in a serious conflict between a "categorically objective" value of information and the ethics of his social function, which is to help the patient to the best of his or her ability, and accumulated expertise of his professional role (the average accepted standards of care). That there is no such "categorically objective" value of information I hope I have made sufficiently clear. The very provision of information can sometimes be an unnecessary intervention.

Besides the criterion of contributing to functioning, there is another essential consideration in determining necessary and unnecessary intervention, and that is compassion. Forcing a person to continue to function in the face of realistic and unremitting aversive environments is not an ethical act. It would be acceptable if ethics involved only a rational program for contributing to functioning (continued survival). But for two reasons, such a program is indefensible:

(1). The universe has characteristics of *both* negative entropy (growth, more organization) and positive entropy (breakdown, disintegration) and we have no *rational* reason for preferring one to the other, only basic affects of yea-saying or nay-saying.[24]

(2) Because we are a social species, we also have social feelings of response to distress, care and concern, empathy.

These two affective considerations are basic to ethics, more primary than considerations of survival and well-functioning. Hence, we are not in an unlimited interventionist program, nor has medicine historically been a "life at all cost" proposition. As a result, intervention must be evaluated in terms of the probability that it will restore some minimum level of functioning (quality of life), and that level must be one the patient would choose to live with as well [e.g., the Quinlan case, In re Quinlan 70 N.J. 10 (1976); Saikewicz vs Supt. Belchertown State School 370 N.E. (Mass.) 2d 417; Dinnerstein]. The physician, in terms of caring, may even recommend against an intervention, as is done in many dialysis

[24]Colleen Clements, "Stasis: the Unnatural Value," *Ethics* **86**, 136 (1976); and "The Ethics of Not Being: Individual Options for Suicide," in *Suicide: Contemporary Philosophical Perspectives*, St. Martin's Press, New York, 1980, pp. 104–114.

units. This human response of compassion can and should override the value of survival at all cost, for the philosophic reasons I have advanced. We lose sight of this when we try to force ethics into a rationalistic model and forget the first emotive premises of any axiology. Fletcher and Taylor, from their different perspectives, both achieve the same humanizing effect in ethics by pointing out the basic affective component (for Fletcher, *agape*[25]; for Taylor, compassion[26]).

Catch-All Explanations

Case Study 19. Possible Ataxic Cerebral Palsy

The patient is a 4-year-old girl whose parents came in for counseling about future pregnancies. The child was slow in walking, has reduced reflexes, and limited, impeded speech. Although the pregnancy and delivery were normal, the first diagnosis was ataxic cerebral palsy. An uncle's daughter is retarded. The patient was taken to see the neurologist treating the uncle's child. A diagnosis of possible tuberous sclerosis (TS) was made. A dermatologist disagreed. A CAT scan was negative. Tuberous sclerosis is now unlikely (if it were the diagnosis, there would be a 50% recurrence risk). Although the parents had the impression that the TS diagnosis was firm, the neurologist's letter to the genetic counselor was much more guarded, and a second letter agreed that the diagnosis was unlikely. The etiology is now unknown.

It is often difficult for a physician to admit that the etiology of a disease or impairment is unknown. Rather than face this unpleasant fact, and in the face of the patient's need to label or identify the problem in some way, catch-all explanations are frequently used. A "cerebral palsy" diagnosis can often be a dumping ground for unknown etiologies. This spuriously fulfills two needs, preserving the physician's role as expert when that role is erroneously perceived as requiring the expert to know all the answers, and supplying the patient's need for identifying in some way what is wrong. This identification ideally allows the patient to work out a problem because it is now a known factor, one that can somehow

[25]Joseph Fletcher, *Situation Ethics: The New Morality*, The Westminster Press, Philadelphia, 1966.
[26]Richard Taylor, *Good and Evil*, Macmillan, New York, 1970.

be controlled by learning about it, incorporating the information into oneself. This can be a very useful adjustment mechanism. However, catch-all explanations can fail on both these grounds because they are inadequate. If they mask the real diagnosis, one that may erupt with all flags flying, the physician may look not like the expert, but a fool for missing the diagnosis originally. The patient, through paying careful attention to his or her own body, may discover that the labels really say nothing and that the need for understanding is not actually being met. The possibility of and responsibility for error needs to be analyzed.

Case Study 20. Cerebral Palsy with Autism

The wife had four children by a previous marriage, one of whom was severely mentally retarded. He is 18-years-old, with a diagnosis of cerebral palsy with autism. The wife was very anxious about the risk of another retarded child, although the husband was not as fearful. The amniocentesis test was explained as not guaranteeing a normal child or capable of eliminating all fears. In this case, estimating recurrence risks is difficult because of the already previously mentioned catch-all nature of the diagnosis, compounded by an autism diagnosis. The couple is beginning to appreciate the imprecision of such diagnoses since estimating recurrence risks demands more precision than such diagnoses can achieve. In order to relieve their anxiety about the nature of the boy's retardation, more information is needed than can be supplied by such an explanation, and they are beginning to realize this. The label is failing to serve its legitimizing purpose. Autism diagnoses, in addition, have historically been characterized by errors. Many currently labeled autistic children were previously diagnosed as mentally retarded or psychotic, and institutionalized without the treatment called for by a diagnosis of autism. In fact, autism represents a relatively new disease category, a new way of clustering symptoms from the general sea of symptoms. Sometimes the problems of theoretic categorizing (creating a consistent taxonomy) become so absorbing and complicated that they negate the pragmatic value of such categories. Some categories may require empirically nonfeasible precision (as in psychiatric nomenclature[26]), while others are too imprecise to be of much help (cere-

---

[26]C. H. Ward, A. T. Beck, M. Mendelson, J. E. Mock, and J. K. Erbaugh, "The Psychiatric Nomenclature," *Contemporary Readings in Psychopathology*, John M. Neale, Gerald C. Davison, and Kenneth P. Price, eds., Wiley: New York, 1974, pp. 20–28.

bral palsy, failure to thrive, mental retardation). Should a physician use either kind of category knowing that the primary effect will be to enhance his role as expert?

Case Study 21. Developmental Delay

The young mother had been sent for chromosome analysis on her baby to eliminate the possibility of Down's syndrome. There was no indication on the physical exam for Down's syndrome and therefore a test would have been a superfluous intervention. There was evidence of failure to thrive, but a careful interview and gathering of information indicated that the cause was psychosocial: mothering deprivation. The mother was herself under considerable stress. Nevertheless, the baby did well because the grandmother was supplying maternal care. When she was absent, the child's growth slowed dramatically. On her return, the child again thrived. This not only illustrates the psychosocial aspects of medical practice, but indicates how very little information labels like developmental delay or failure to thrive give us.

Gorovitz and MacIntyre nicely analyzed some of the possibilities for error in medicine in a discussion relevant to the issue of the possibility and responsibility for error that these case studies point to.[27] They made a distinction between the internal norms of medicine as a science and the external norms that are motivations for entering and practicing the science as characterizing present assumptions about medical error. The only sources of error, then, are ignorance (which is an integral part of scientific progress) and ineptitude. The example they give is a physician-prescribed drug with unfortunate side effects for the patient. Error is the result either of physiological and pharmaceutical ignorance, or the negligence of the physician. They propose (in serviceable systems theory concepts, but not systems terminology) that a third source of error is present: when we attempt to explain a concrete particular by generalization, we are using a generalization based on the importance of similarities and not differences among units. But differences, on the concrete individual level, can be important. In systems language what they are saying is that error has a major source in the confusion or mixing of levels of organization, levels of complexity, levels of perspective. On the functioning individual level, my previous discussion of the unfragmented, whole patient

[27]Samuel Gorovitz and Alasdair MacIntyre, "Toward a Theory of Medical Fallibility: Distinguishing Culpability from Necessary Error," *Hastings Center Report* **5**, (Dec., 1975) 13–23 (1975).

dovetails with their emphasis that these particular individuals must be understood as self-maintaining wholes functioning or malfunctioning in a changing environment. The old axiomatic model would have to be expanded (probably in terms of detailed and ongoing expansion of initial conditions) to such an extent as to be impractical. They point out that it would also result in inevitable predictive failures. The basic reason for this is the poor fit of the old model to process and change. Warranted assertability (truth, if you wish) is a moveable measure.

The authors seem to feel this would be news to physicians, and that the general public also has some sort of false confidence that their doctor will not make a mistake. Gorovitz and MacIntyre want the physician/patient relationship redefined as a relationship in which mistakes are necessarily inherent, on the basis of the three reasons for error they have given. Although it is true that physicians wish their patients to have optimistic belief in their ability to help for good biopsychosocial reasons, the doctor's awareness of the possibility of error is always strong. The physician may need to feel that in this particular case a mistake will not be made; nevertheless, that doctor always knows a mistake can be made. The hope is that if such an error is made, it will not be an irrevocable one. Even common folklore, however, acknowledges that such mistakes can indeed be irrevocable: Doctors bury their mistakes.

The empirical nature of medical practice makes error a built-in feedback. The actual treatment of a patient is not the collection of sophisticated test results, the correlation of the results with a law-like diagnosis, and the unswerving application of the generalization in the form of unchanging treatment. It is a much more jerry-rigged affair, truly empirical, and constantly dependent on feedback. The doctor tries something. It does not work, and is modified or replaced by something else. The response looks good, so the physician goes a bit further until finally satisfied that the diagnosis is correct. It might have been the one that had first come to mind, or it might be something very different, but the process is always tentative and probing. In a sense, it is finally biopsy or autopsy that removes most possibility of error.

My guess is that the patient knows all this, at least when it does not seem personally necessary to practice denial. The myth of medical infallibility may be a medical ethicist's myth, but it is not a cultural one. The patient does know that the doctor will do something (even if it is only supportive and palliative; or worse, wrong), and that may be the important element. The patient knows someone is trying. Recalling my discussion of the two ways of viewing pain and suffering, the patient, by going to the physician, has made a choice between the two ways of coping—that is, the interventionist alternative of doing something, or trying to do something, about reality has been chosen.

The enormity of the responsibility assumed by the professional, in the light of the constant possibility of error, is as awesome as Cassell paints it.[28] Denying that by emphasizing the physician's role as a specialist or technician may be the defense mechanism Cassell thinks it is. There *is* the problem of professional burn-out, although such a problem is hardly unique to medicine. Any social role that impacts directly and with high visibility on individuals' lives is prone to it. This responsibility stems from the basic attitude of compassion and the desire to help. It is the arational base for the medical role and the justification for assuming this responsibility in the first place. Given the desire to help, one can proceed rationally or irrationally. When the approach is irrational and produces consequences that add to suffering, when the error might have reasonably been prevented or not being foreseen might still have been remedied but was not, then responsibility becomes culpability. This view of culpability is broader than the legal one and based not on the concept of punishment but on that of shame, what L. Jolyon West has described as a much more powerful aversive conditioner. Professional standards, when internalized, are potent reinforcers and a preventive safeguard rather than a punitive safeguard. No one is terribly happy with the concept of defensive medicine. The adversary aspect of the physician/patient legal model also seriously modifies the notion of responsibility. Responsibility becomes no longer an ethical concept, but a by-the-book notion of contractual obligations. The physician's existential self is no longer threatened by such responsibility, only the financial self. Both models could perhaps co-exist. I believe I know which one is essential.

New Syndromes

Study 22. Hyperpigmentation

The mother brought in her two young sons to be checked for excessive pigmentation. The one boy primarily has excessive pigmentation of the nape of neck that appears innocuous. His brother, however, has more extensive capillary hemangioma on

[28]Eric J. Cassell, "Making and Escaping Moral Decisions," *Hastings Center Studies*, 1 (No. 2), (1973).

the entire right arm, chest, and leg. On the left side, although hyperpigmentation was not present at birth, it is expanding now. The mother would like to know if this is heritable. The boys appear healthy on physical exam, with no visual or auditory complaints. The physicians have never seen anything like this. The best preliminary label is: probable unknown sporadic event, which leaves most questions unanswered. Are the two types of lesions related? Is it genetic? What are recurrence risks? What is the prognosis for this hyperpigmentation? Should we worry that it might present problems similar to those of neurofibromatosis? Should we be concerned with the link between pigmentation abnormalities and malignancy? The physicians decided to be as reassuring as possible. They would not convey concerns about malignancy or neurofibromatosis-like problems because there is no information to base these concerns on. To alarm these people without any real knowledge to justify that alarm would be considered professionally irresponsible. A new syndrome, such as this appears to be, creates management problems. Monitoring the condition to try and learn whether it poses a health threat needs to be done without the observation seeming to imply that it is a health threat. Predictive ability is nil; the possibility of error is considerably heightened. In addition, the patient is aware of the physician's lack of knowledge and may realize it is a learning experience for both of them. Does medicine still perform an important function for the patient, despite the lack of knowledge?

Case Study 23. Unreported
Balanced Translocation

The couple has one living child and a history of four spontaneous abortions. A chromosome test showed the mother was a carrier for a balanced translocation not previously reported (involving chromosomes 4 and 7). A very significant amount of chromosome material is involved. Further pregnancies can be monitored with amniocentesis, now that the risk is known. The wife then told the physician that her sister, who was 4 months pregnant, had one child and a history of two spontaneous abortions. It was suggested that she tell her sister this new information, but she preferred that the counselor tell her sister's obstetrician. The obstetrician, after informing the sister, suggested a chromosome analysis for her. The results showed that the sister was also a carrier for the new translocation. Amniocentesis was scheduled for her as soon as possible.

This case represents a new syndrome that offers much more information for the prospective parents and an enhanced ability to deal with the medical situation, something significantly lacking in the previous case. It is peculiar in that we know the genetic mechanism very thoroughly, but now do not know the resulting syndrome's symptoms, other than it may be a lethal. Thus far, fetal loss is the known expression. This lethal factor is consistent with the amount of chromosome material involved (a major impact on development is to be expected). In this case, the appearance on a cytogenetic level of a new syndrome has actually supplied more options, more choices for the patient and the patient's extended family, tending to verify Lewis Thomas' observation that when we really understand what's going on, our responses tend to be simpler and more effective.[29] I will discuss how such new information could affect self-image and the ethical conflicts that involves in another chapter.

The final point I wish to make now about the value of information and what its proper role is in medicine stems from the previous case. Is the generation and imparting of information, and manipulation as a result of this, the only important function of medicine? Is it the overriding value in the medical enterprise? The view of medicine as technology, or as a science of biological research and technological intervention, would give an affirmative answer to those questions. It would not be an accurate description of what physicians actually do in the practice of their profession, of course, but we must consider whether it represents what physicians should do.

For that to be the case, information must be seen as an intrinsic value, as I have previously discussed, or as the highest ranking value in a hierarchy. My analysis indicates that it is neither. Research and intervention, then, are part of a much broader interaction, and not always the most important part. Medicine is generally viewed as a helping profession, which should give us a clue to what the other components are. It involves a personal interaction often centered on those primary motivators of human behavior, pain and fear. Although it is possible to try to restrict the interaction to information conveyance and technical manipulation (which would considerably lessen the traditional burden of responsibility for physicians), such a move demands more of information as a value than it can produce. Information may assist us, but is not adequate to enable us to cope with pain, fear (anxiety),

---

[29]Lewis Thomas, "The Future Impact of Science and Technology on Medicine," *BioScience* **24** (No. 2), (1974).

and concerns about death. Religious help assumes the acceptance of these realities, but as I established in the first chapter, it is one of *two* major ways of dealing with them. The patient is usually motivated by the second attitude, a desire to attempt to change an unacceptable reality. Intervention is desired, but a broader intervention than the medicine-as-technology model, since the science must be related to the pain, anxiety, and mortality to fulfill the desire. That relating involves the physician in an interaction with the patient that is total, that is not limited to the value of information, but addresses other values (interests) of the patient. Even in the brief encounter that genetic counseling often is, those considerations are taken into account.

I fail to see any other way the physician can help in a manner that promotes the best interest of the patient. That best interest is determined by the personal interaction that occurs, and varies from specialty to specialty, but is not the one-way street that the paternalism objection suggests. Neither, however, is it any or all of the technocrat, mechanic, or party-to-a-contract models suggested as alternatives. These do not sufficiently address the full range of relevant interests the patient has in coming to the physician, and operate on the incomplete assumption that the physician is restricted to information as a value in itself.

Summary

In this chapter, we have investigated how we make the choice for information, and have found that the choice is made relative to other values and that the process of choosing defines the sense of the various options. In addition, we have seen that the ranking of values is hypothetically, rather than categorically, objective. The value of information is thus based on the purpose we have in mind: we always need to ask, information for what? The fundamental answer is always that compassion or the wish to relieve suffering constructs the purposes for both patient and physician. There are general implications of such a value theory. Values are seen as choices, ultimately, but not as exclusively rational choices or choices determined by rational principles. Affective components, such as purpose, interest, social bonding, and an affirmative attitude toward existence, are therefore seen to be major elements of any viable ethical system. Finally, applied ethics can now be described as a system in which there is a changing ranking of multiple values and in which an attempt is made to maintain a dynamic balance among them.

Chapter 3

Denial and Reality Testing

In the last chapter, I introduced some of the ethical considerations involved in the common human response to medical information—denial and reality testing. In this chapter, I would like to go into this in more depth; but first, an explanation is probably owed the reader who is wondering why the concentration on such issues when most of the literature of medical ethics deals with abortion, passive euthanasia, compulsory sterilization, compulsory screening, and eugenics. My experience in a genetics clinic has been that such issues are not the real issues that arise with any degree of frequency in such a medical program. The routine ethical concerns instead center around such management issues as denial and reality testing, and around the value to the patient's life of the genetic information available. I will discuss abortion as one of the options available in reproductive decision-making in some detail in a later chapter, and will touch on each of the other issues to some extent. But the bread-and-butter medical ethics issues, as they present themselves in actual cases, are in a sense more mundane, and in another sense, much more significant, than this. They consist of the physician/patient interaction and the justification for the routine interventions comprising that interaction. They are less glamorous, but decidedly more human.

Underlying all our concerns about the choices between denial and reality testing is our basic evaluation of anxiety: our perception of the part in normal living played by anxiety, how "not-anxious" is it possible for a human being to be, and whether we would want a human being to be that tranquil. The case studies that follow, as well as some of the discussions in the literature, underline the importance of understanding just how we view anxiety and how we perhaps should view it. Physicians are very concerned about elevating their patients' anxiety levels. We need to take a look at the role anxiety and its induced behaviors play in living, and what value or disvalue we want to assign, in a systems concept, to the raising or lowering of this affect.

Medicine tends to adopt the same approach to anxiety as it does toward pain and suffering, with the same problems arising. The difficulty is to maintain the delicate balance between present anxiety and its long-term beneficial consequences, between the prevention of anxiety and the possible increase in future anxiety that such prevention might entail. Anxiety, viewed as psychic pain, tends to be dealt with in much the same way that medicine deals with physiological pain. However, the concept of psychic pain is even more complex than physical pain, even more dependent, because of the lack of understanding of as many of the physical correlates, on reasonably delineated contexts. Again, a look at some case studies can help adequately draw those contexts.

How Much Patient Denial Can Be Accepted?

Case Study 24. Meckel's Syndrome

A pregnant woman, over 40, came for counseling because her first baby died within an hour of birth. Examination had revealed polycystic kidneys, neural tube defects, and a cleft palate. A rather rare autosomal recessive disease, Meckel's syndrome, was diagnosed. It is usually rapidly fatal after birth, because of very poorly functioning polycystic kidneys and significant brain abnormality. An amniocentesis was done, but the cells failed to grow, the first culture failure for the lab. The first alphafetoprotein (AFP) test gave a borderline result, but the level was extremely high on a second test. Another amniocentesis was performed during the woman's nineteenth week of pregnancy. At this point, there was accumulating evidence that the lethal or sublethal abnormality would be found. Cell growth for the second tap again was peculiar. There was half-hearted consideration of amniography, a more invasive X-ray procedure requiring the injection of dye, but the benefits seemed slight in comparison to the risk. The genetic counselor was, however, reluctant to interpret the AFP levels, poor culture growth, and previous history in a very negative way, and wanted, for this case, clear empirical evidence before stating that the fetus had a very high probability of suffering from Meckel's syndrome.

An ultrasound scan was done during the next few days. This technique is becoming much more precise, yielding refined fetal visualizations. In this case, it showed a definite encephalocele. A complicating factor arose; the woman was found to have cervical carcinoma *in situ*. With the baby projected to die within a few days of birth, and with the new health problem for the mother, her physician wanted to do a total hysterectomy. How this woman hand-led the situation will give us some idea of the importance of the patient's denial or reality testing behavior in medical deci-sion-making. In spite of the now uncontested diagnosis, the par-ents decided against an abortion/hysterectomy. Three reasons were given. They wished to maintain reproductive capacity (in spite of the information on carcinoma and Meckel's syndrome); they said they were religiously opposed to abortion (we will see later that the woman still thought the baby might be a normal baby); and they wanted no further "doctor management" of the problem. Contact with the physician became sporadic (missed appointments).

The patient delivered a baby who lived 15 minutes. Autopsy revealed an encephalocele, polycystic kidneys, liver cysts, poly-dactyly, and poor eye formation. The mother had denied this risk to the last moment, talking instead about whether she should breast feed or bottle feed the baby.

It is possible that the genetic counselor's desire for complete verification of the Meckel's syndrome diagnosis for the fetus may have been a response to the patient's wish to deny the possibility of a seriously defective baby. Patient denial, in the physician/ pa-tient interaction, will thus often have some effect on the physi-cian's decisions as well. The efforts of the physicians to present the reality of the situation to the patient were seen as unwarranted medical management. How directive should the physicians have been; how much denial should be allowed? If the couple still wished to reproduce, is this self-destructive behavior and are there ethically justifiable limits to the social tolerance of such behavior? In this case, anxiety was raised by the medical interaction and could not be dealt with by the patient, resulting in hostility toward the bearers of the news and refusal to accept the information (real-ity). Did this actually benefit the patient in the short-run, or were both short-run and long-run consequences deleterious? Could fa-cing the reality of the situation have changed it in any beneficial ways? If anxiety does not lead to long-range benefits, is the physi-cian justified in raising it? Does it make any sense to say that anxi-ety would not or could not lead to long-term good? Is this case such a situation, or what would such a situation be?

Case Study 25. Fetoscopy for Thalassemia

A couple with two children affected by thalassemia accidentally conceived, in spite of contraceptive measures. Because of the ill-ness of their two children, they were firm about not having another child with thalassemia. Prenatal testing for this problem is at this time experimental, and carries an unknown risk. Fetoscopy is necessary to visualize and draw blood from a placental vessel (new recombinant DNA techniques may make detection much less intrusive). Without testing, the couple had decided they would abort. The test was done and indicated the fetus was heterozygous, so the pregnancy was carried to term and a baby boy delivered. A week after delivery, the pediatrician and genetic counselors discussed whether a routine blood sample should be done on the baby. The blood test would confirm the prenatal diagnosis (it is not known what the accuracy of the prenatal test is because of the small number done). The laboratory is, of course, interested in quality control on this experimental procedure. The blood sampling is not invasive. The parents have not requested confirmation, however, and the pediatrician does not want to bring it up because he fears it will raise their anxiety level. The physicians involved all strongly wished this couple's baby would be healthy. He certainly seems to be. Should everyone wait? Are the parents already anxious, since they were carefully told about the inaccuracy factor in the prenatal test, and would a blood test now on the baby alleviate more anxiety than it produces? Since heterozygotes are sometimes more difficult to separate from affecteds than homozygotes are, should this make a difference? Could the doctors' concern about raising anxiety levels also be a denial mechanism on their part? Are they anxious about the possibility of having to tell the parents the prenatal test was wrong? Denial, like intervention, sometimes requires collaboration. If the physician does collaborate in denial, what responsibilities are ethically assumed? And exactly how does the physician predict that anxiety will be raised?

Case Study 26. Previous Anencephaly

The couple have two normal children, and have also had an anencephalic baby. After this experience, they did a great deal of reading about neural tube defects. Although the husband would want to abort if NTD were diagnosed, the wife is uneasy about abortion. She realized that having two normal children was no reassurance it would not happen again. Their perception of the reality of the situation is good. An alphafetoprotein test can help, and they elected to have the test. Ultrasound can also give them needed information. In this case, tests are available to the couple. The problem of choosing between abortion and a baby with NTD remains, however, if the tests were positive. Has this couple's choice for reality testing given them more options? Reality testing is another way of dealing with anxiety. Is it more suitable than denial in certain situations? In all situations? Is it only functional when intervention remains a possibility?

Case Study 27. Two Spontaneous Abortions

The couple was seen 10 days after the second abortion (first trimester). There is a very strong drive on the wife's part to become pregnant. After the first abortion, she went through a very thorough workup and seems almost to be demanding an answer why. She impressed the physician as rigidly trying to maintain control of her body processes. During counseling, she expressed concern about a psychosomatic cause for the abortions. However, although this hinted at an intrapsychic conflict, she sealed herself off from the physician and a good interaction did not occur. Chromosome analysis was suggested in a reassuring way, but she was not reassured.

Although the patient initially seemed to be engaged in reality testing, a closer analysis indicates denial. Her real concern, her real problem, was not being expressed and she resisted interaction with the physician on this level. A problem for more directive counseling is precisely this withdrawal and resistance. For direction to reflect the patient's best interests, the physician would need to know the patient well enough to realize what those interests are. When the patient denies the problem in the interaction, this is usually what does not occur. This makes it very difficult to evaluate the patient's denial need, to ascertain how much anxiety will be produced and what coping mechanisms for that anxiety exist, and to project what the long-term effects will be. Should all denial, then, be allowed? Should the first level of resistance be the limit for medical intervention? One of the prices of denial is that it cuts us off from many human interactions and relationships. This is also one of the results of excessive concern about raising anxiety levels.

Case Study 28. Maternal Age

This seemed like a straightforward, maternal age counseling problem. The woman was almost 35 and had been referred by her obstetrician, who in fact called for an appointment. She had not been concerned about abnormalities, was Catholic, and would have a problem with abortion. She *was* concerned with any test risk for the baby. She delayed seeing the counselor and after counseling, had not made a decision on having the test. She had heard about lawsuits against physicians who did not refer their 35+ patients for amniocentesis and felt very strongly that physicians should not be legally culpable. What she was really saying was that her doctor should not be legally at risk for not referring her, and she would have preferred not to see the genetic counselors. She wanted to be able to deny the 35+ risk, but the legal considerations created a situation where denial was difficult to maintain. Did the legal precedent and the doctor's referral, certainly an instance of defensive medicine, violate some "right" of denial for her? Would this be a valid ethical objection to the legal decisions that have required doctors to inform patients about the availability of amniocentesis?

Case Study 29. Maternal Age

This couple is very anxious. Their obstetrician was quite directive about the test, but they really do not want to know whether an abnormality is present. The father was very concerned about needle damage to the baby, but the main concern was to avoid any stressful information. The counselor discovered that this couple had the perception that no one liked them and that their personal life situation was very tense. They probably cannot deal with the pressure of a test. Is some denial necessary? Is this a no-win situation: if they cannot deal with the pressure of a test, could they deal with an abnormal baby?

There is of course straightforward denial and devious denial. Scrambling and forgetting information is common. Seeming to misperceive reality, while really not so doing, is also patient behavior. The ethics of the profession itself puts some limits to denial. NIH, for example, has standards on what information must be conveyed to the patient, depending on risk levels.[30] As I discussed in the last chapter, the patient does not have an unqualified "right" to decide how much information to absorb. The first limit to denial is based on the expert medical role: there is a lower level of information that must be conveyed and understood if an amniocentesis test, for example, is to be done. But this is a very low-level competency requirement, as usually interpreted, and still allows a great amount of denial.

A second level is that of mild discomfort and resistance on the patient's part. To draw the line here would imply that existence

[30]USDHEW, NIH, *Institutional Guide to DHEW Policy on Protection of Human Subjects*, Dec. 1, 1971.

should never be uncomfortable, that anxiety cannot be handled, that beneficial growth does not occur through mildly painful sensations. This view forgets that although pain (physiological or psychic) is an intrinsic evil, it is often an instrumental good, and crucially so as a motivation for adaptation and change. Such a view also puts the physician in the position of being so protective of patients that responsibility and intervention must be unrealistically overextended, or else must be restricted very sharply indeed.

On this level, although the patient may attempt some denial in the interaction, the physician, a part of the process, can continue to give information and support. If the physician draws back too quickly from this encounter, then that doctor's own reality testing, attitudes toward anxiety, risk, and the possibilities of human behavior need to be analyzed.

Levels of denial escalate, however, as the case studies illustrate. The ethical basis for challenging denial needs to be well-spelled out on higher levels. We may have a case, on occasion, for paternalism. However, such paternalism could result in the decision to break through denial defenses or to leave those defenses in place, dependent on the ethical justification for challenging such a denial in the first place. Let me first outline what I think such an argument or justification would involve, and then more closely examine case studies involving questions of raising anxiety, since an analysis of anxiety-in-actual-context is crucial to this whole question.

(1) Anxiety can serve either as a motivation for change and problem-solving or to generate a defensive block that perpetuates the current conditions for as long as possible. It is first necessary to determine which function anxiety is performing, or could perform. This is a judgment that can be made only by knowing some important aspects of the patient's personality and life events. The actual practice of medicine makes the assumption that the physician/patient interaction yields this information. From a study I recently completed on practicing physicians in a seven county area, my in-depth interviews indicated that physicians felt this assumption was valid.[31] In a genetic counseling situation, the briefness of the contact might not generate sufficient information.

[31]Sorensen, Parker, and Clements, "A Study of Foreign Medical School Graduates in the 7th Medical District," 1978.

(2) If anxiety results in a defense, denial, two things are again going on. The psychic pain is repressed or avoided, and the status quo, and hence the problem, continues. It is now necessary to determine whether insisting that this pain be acknowledged and experienced has more value for its instrumental result of beginning work on the problem than the intrinsic disvalue of the anxiety.

(3) There are four possible combinations:

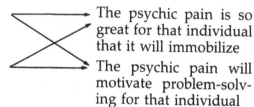

The problem can not be solved or much alleviated → The psychic pain is so great for that individual that it will immobilize

The problem can be solved or alleviated → The psychic pain will motivate problem-solving for that individual

Each of these combinations occurs in a context of varying degrees of severity of the problem, so that the projected consequences of not solving or alleviating the problem need to be variously ranked on a scale of bad consequences. These combinations also occur in a context of individual response to psychic pain, which widely varies depending on previous life experiences and current support systems. A consequential calculus is not all that easy to construct. It depends on the premise that anxiety is a possible instrumental good while an intrinsic evil, and that denial is instrumentally either good or bad, but has no intrinsic value. It is a response to anxiety, which may either increase or decrease pain. It is a flight (or perhaps better, a freeze) mechanism that a system can call on in stress situations. To label it better or worse than a fight (or coping) mechanism makes no sense, since the value of both depends on the context and outcomes they generate. If, for example, two of the combinations are considered, denial may serve a good purpose: (1) The problem cannot be solved or much alleviated, and the psychic pain is so great for that individual that it will immobilize, and (2) The problem cannot be solved or much alleviated, and the psychic pain will motivate problem-solving.

There are two things to consider here. First, recall my discussion of the difficulties in determining when a situation is hopeless. It is not a cut-and-dried task and we do not want to close out our options. Secondly, denial can cut us off from many social interactions that might, at least internally, alleviate the problem. A socially caring environment can constitute such a solution.

There are, then, some cases where denial might be an instrumental good, even though the cost of denial is very high. Medical practice, usually openly interventionist, may be overly eager to embrace these cases as typical, since intervention on a psychological level is much more hesitantly acknowledged. The primary expression of this is the doctor's concern about making the patient more anxious, so that I propose now to take a closer look at anxiety-raising.

Anxiety Induction and Lowering

Case Study 30. Marfan's Syndrome

Counseling is being given to a family who lost a 20-year-old son because of an aortic aneurism. An autopsy was refused, but there is suspicion of possible Marfan's syndrome. His mother's sister has a child with this problem, and the paternal grandfather died as a result of an abdominal aneurism and stroke. His father has cataracts and all of the family are very tall (there are six brothers and sisters). There is no definitive test or effective treatment for Marfan's syndrome, so a question arose about how much of the extended family should be contacted? If we raise their anxiety levels, is there any way to work toward long-range solutions or a lowering of that anxiety? Since the siblings are at reproductive age and Marfan's syndrome is dominant, would short-range increased anxiety outbalance the psychic pain resulting from the birth of children with the genetic disease?

Case Study 31. Gonadal Dysgenesis

A 24-year-old, engaged woman came for testing and counseling. Her medical chart indicated that after birth, her sexual designation had alternated. Physical exams showed a large clitoris, a "normal" uterus, and ovaries. Turner's syndrome was diagnosed, but surgery a week before her visit to the genetic counselor had indicated, along with a possible Fallopian tube, also "possible testes" tissue. The girl is short and has amenorrhea. There is a possibility of mosaicism and a possibility that the identification of "testes" tissue is inaccurate. If this incomplete information is presented to her, her sexual identity will once more be in question, and she has been functioning as a woman. Her anxiety will be raised, and to what purpose? Is this a case for withholding information?

Case Study 32. Possible Down's Syndrome

The mother had amniocentesis, which indicated a normal child. The baby's pediatrician, however, has asked the lab if the test might have been inaccurate (except for mosaicism, probably not). He finds the baby to have slightly abnormal eyes and hypertonia. Should he pursue his question with the parents? The genetic counselor thought not. It would raise the parents' anxiety and very probably change their perception of their baby in a negative way. Reassurance and the lowering of that anxiety might not be achievable, since the possibility of mosaicism could never be ruled out completely. The pediatrician first needs to make sure there is sufficient medical evidence of a problem. Is it present? What constitutes a problem?

Case Study 33. Juvenile Onset Diabetes

The first pregnancy had ended in a stillbirth. Autopsy revealed the only abnormality to be one overly large, and one collapsed, lung. The mother is very anxious and alarmed over this; the father is not interacting. Family history indicates a cardiac defect on the father's side and diabetes on the mother's side. The very anxious mother does not want a child with diabetes or cardiac defect. The counselor was reassuring about both. Actually, the diabetes risk is slightly higher than background risk, but since the couple was so anxious, the counselor did not go into this in any detail. The mother's perception of the problem seemed exaggerated, and the counselor wondered if something in the family history was the reason for this. During the session, her hands were ice-cold; she was extremely anxious. A followup was felt necessary. Is this abnormal anxiety? Are individual coping mechanisms too weak to allow detailed presentation of the problem?

Case Study 34. Encephalocele

The couple's first child had an encephalocele that had been corrected surgically with no apparent adverse effects. Because the woman was pregnant again, their obstetrician had insisted on referring them for counseling. Before the session, however, they had made up their minds not to have amniocentesis, for a number of reasons. They are Catholic and opposed to abortion; he is a math teacher and suspicious of statistics; she is not sure how she could use the information. They felt the genetic counselor was very directive, although the counselor attempts to be nondirective. The wife was using a great deal of denial. The husband had interpreted the neurosurgeon as being very positive about the problem and as informing him that the cause of the defect was unknown and nongenetic. He felt the counselor was emphasizing a genetic component and the negative possibilities, and was resistant to this. The genetic counselor here was a provider for an unwilling consumer. Raising this couple's anxiety level might generate a hostile reaction. Should the obstetrician have referred them at all? Was the neurosurgeon as positive as reported? Was the genetic counselor as negative as perceived?

Case Study 35. Diaphragmatic Hernia

The patient is a very anxious woman who missed three appointments. She has a son with a diaphragmatic hernia and told the physician that if she gave birth to another defective child she would kill herself. There is a family history of diaphragmatic hernias on her mother's side and of possible Down's syndrome on her father's side. Her mother will not release information about her sister, who died from the defect. The patient's reaction is extreme. Family dynamics do not appear supportive. The missed appointments may indicate that she is operating close to the limit of her personal anxiety tolerance. She was told the ordinary risk is 1 in 50. Should she be further referred for psychological support? Is the risk figure all she needs, and will it lower or increase her anxiety?

Case Study 36. Anesthesia Risk

This patient had a very low level of anxiety, as opposed to the previous cases. But the details of her case would ordinarily be very anxiety-producing. She has two children by a previous marriage, and had separated from her second husband after being married a year. He is currently institutionalized for drug and alcohol abuse. Before institutionalization, however, he forcefully impregnated her to prove his fertility (and probably masculinity). About the same time, she had an operation with general anesthesia. If her reported conception dates are right, it was safe. If her obstetrician's estimate is right, it was not safe. Ultrasound can establish which is probably correct. She wants to put the child up for adoption and would abort if it were abnormal. Need the physician worry about raising her anxiety level in terms of her apparent coping capability? Is the capability only apparent?

Case Study 37. Down's Syndrome and Rubella

This is the wife's first pregnancy. She is a social worker who feels very strongly about not having a defective child, and this will be their only child. Her anxiety level is very high, but indications for it are minor. Her uncle's child might have Down's syndrome. She also knows that working with children might have brought her in contact with rubella (German Measles), but the real risk period for rubella is over. The family history she gave the obstetrician varies from that she gave the counselor. The counselor felt she might need additional counseling and also that she qualified for amnio-centesis on psychological grounds. Her anxiety is so high that the test could serve as a means to lower it, even though other indications are not strong for performing it. Medical intervention can lower anxiety levels. Is this a valid use of medical technologies?

Case Study 38. Mucopolysaccharidosis, San Filippo

Tests were suggested to establish the diagnosis of an institutionalized son. No procreative questions were concerned. His mother has decided to have no more children and no other relative is involved. What is the value of diagnostic tests in terms of anxiety?

I have previously discussed the anxiety lowering benefits of information, of naming the problem. This can be a working-out process and be of considerable psychological comfort to the family.

Case Study 39. Maternal Age

This was routine maternal age counseling. However, the couple expressed anxiety at first because the physician had brought up a risk they were unaware of before. They left, nevertheless, feeling that this information should be made available generally.

In most interactions, the raising of anxiety can be dealt with and can serve as a preventive measure to greater anxiety in the future.

What these cases illustrate, and I have included a great many, is the wide range of responses and situations in which anxiety is a factor. Although it is only one of the factors, since its potential for motivating behavior is very high, it figures very importantly in decision-making. What is needed, however, is to discuss such considerations in a realistic context.

To remove anxiety from the world might not be the imagined good we would like to think it is. Like physical pain, anxiety is a warning signal to an organism that something is wrong or needs changing. Since we do not live in a static universe, our balancing act in terms of our changing surroundings needs to be continuous. To keep a workable balance, we need feedback that informs us when that balance is threatened. The distress mobilizes activity to lessen the distress, which is a normal characteristic of living, if living is a process, an interaction.

In addition, social interaction implies the potential to create anxiety. Because we are individuals, our needs and interests vary, and the mere act of reporting back from our social environment that the balance is jeopardized causes distress; it also results in attempts to restore the lost personal balance. This is really what we mean when we talk about social characteristics. Someone wants something from another individual, or someone has a want or interest that will in some way diminish that individual. The reporting back from that other individual can be negative and raise anxiety, but unless the reported knowledge can be accepted and adapted to by some change of activity or revised perception of interests, an individual's social nature cannot function. Then we are really left with the Hobbesian view—the war of all against all. But such a Hobbesian description is not the description of a social system as it functions for any animal species I know of that can be remotely described as a social species. "Hurting" (raising anxiety or frustration levels) seems an integral part of any actual relating process, and therefore cannot be avoided if any social relationships are maintained. Without it, our separate interests would continue to collide disruptively, with push coming to shove in an asocial breakdown. Society without anxiety may not be possible.

On every level, then, distress (anxiety) is a necessary evil. Medicine is involved in the same conundrum the universe is: the desire to prevent immediate pain and suffering, and the need to use pain and suffering as instrumental to long-range prevention. To concentrate on one or the other aspect is to distort the picture. It is more risky for the intervener, of course, to raise current anxiety levels than to enter into collusion with the patient's denial in the short-run. The medical expert role and its ethics, however, are based on the potential that the patient will suffer long-range disasters under such a policy of collusion. Granting individual variation, most patients can cope with the anxiety that a "cure" necessitates. The alternatives for those who cannot are no more pleasant to contemplate; their options are severely restricted.

Anxiety, then, is disvalued by human beings in an *immediate* context, which is what we actually mean by calling it *intrinsically* evil. It is a feeling that spurs change and activity, that we choose not to continue to suffer (unless we imagine punishing ourselves will avoid an even greater anxiety). However, in terms of the entire process of living, it is an important instrumental good, leading to adaptive change, problem-solving growth, or at least attempts at these.

There are two foci here: Short and long term. If we slice experience into a long-term segment, anxiety, as part of that segment, is also part of the intrinsic good of that segment, but only as a participating element of the process, not alone. If we slice experience into a short-term segment so short it involves only the present anxiety, then that anxiety is an intrinsic evil (we wish to avoid it). When we try to combine the two ways of organizing experience, mixing levels again, we get the intrinsic/instrumental distinction. We need to remember Sartre's important point that human beings project future possibilities, so experience is a moving, temporal extension, and we can contract or expand our attention to it.

Medical intervention, in seeking to prevent unnecessary pain and suffering, needs to take into consideration the full context to determine reasonably what is necessary and unnecessary. An instrumental good, then, is a choice made by human beings in a long-range context, in spite of the immediate disvalue the choice involves, a matter of the breadth of interest rather than the narrowness of interest; and it seems to me, therefore, not so much a question of different entities, but of changing perspective. Behaviorally, when we do not avoid pain because we view it as a necessary part of a pleasure process, incorporate it into the pleasure segment, we act as if we choose pain (instrumental good). If we mix our perspectives, we can say that there are intrinsic and instrumental goods. It is usually risky, however, to mix perspectives unless we are very careful about what we are doing. Otherwise we're left with these strange creatures, instrumental goods, to try to force into a unified ethical theory. We would then need to explain what could motivate a person to choose an instrumental good in spite of its intrinsic evil, which is not very easily done. This is the old linear cause model, with the problem of the connection between present cause and future effect. In a systems model, where we can look at hierarchies of organization, as it suits our purposes or interests, this problem evaporates. "Instrumental good" is a signal we are interested in a more complex level of or-

Table 1
The Possibilities of Human Attention to Experiencing
and Levels of Ethical Analysis[a]

Level of Analysis 1

Attention to Presently Occuring Sensations
    Interaction with environment causing immediate pain (evil)
    Interaction with environment causing immediate pleasure (good)
    Threshold interactions, neutral until a certain accumulation triggers
    pain or pleasure response
This level has traditionally been described in ethics as intrinsic good or
evil

Level of Analysis 2b	Level of Analysis 2a
Projections of possible experiences	Remembering previous experiences, associating previous experiences
Anticipation of future possibilities can result in our choosing what, on the more restricted attention level, we would choose to avoid (pain, evil) or in our rejecting what we would choose (pleasure, good)	Past sensations can prime our responses and our evaluation of those responses. We can attribute pain or pleasure to this level when no pain or pleasure would be generated by Level 1 alone. Good and evil can be in the eye of the beholder

Level of Analysis 3
Learning and Anticipating

Both learning and anticipating factors render what we choose (good) and
what we avoid (evil) relative to the level of analysis we use. The problem
in ethics is to make clear what level we are operating on, and not move
back and forth without making the change clear. In ethics, the introduction of the concept of instrumental good or evil has traditionally had to fill
this role, but the setting up of an intrinsic/instrumental dichotomy distorts the situation, seeming to create two kinds of good or evil, when in
fact it is the level of analysis that has changed.

[a]Each level contains within it the previous levels of analysis.

ganization, one where the intrinsic good we have chosen can be
broken down into smaller components, some of which may be
called intrinsically evil *only in another perspective*. But we should not
mix the language of the levels of organization to avoid confusing
ourselves. Such terms should indicate when a shift of perspective—a change of the level we are considering—has occurred. We
need to keep these moves in mind since mixing levels really does

result in conceptual confusion, if not chaos, and Table 1 is designed to help us in this effort.

Perceptions of Risk

Risk figures are always presented in rather precise quantitative form, and this is often deceptive. When we ask what the figures mean, the precision evaporates quickly. Since genetic counseling involves considerable use of risk figures, and since they serve as a principal means of reality testing, I want to look closely at this problem.

Case Study 40. Amniocentesis Risk

A question was raised in the clinic by two resident physicians who had participated in counseling sessions for the first time that day. They indicated that they would not do the amniocentesis test unless there was prior agreement by the parents to abort an abnormal fetus, because they perceived the test risk for mortality to be high. The current risk estimate is less than 1 fetal loss in 1000 pregnancies tested, a figure arrived at because the amniocentesis group and the control group of pregnant women not having the test show no statistical difference in fetal wastage. If there *is* a risk signal, it cannot be distinguished from the background noise of normal pregnancy risk, so we use that normal risk figure. Are their perceptions (interpretations of the figure) unrealistic? Recalling my discussion in Chapter 1 of adding "soft" factors to a cost/benefit analysis, are they making narrow protocol decisions about what should be weighed as benefits? Does their perception of risk override the parents' perception? There is an additional factor. This risk figure is based on the entire population of women having amniocenteses. It is not broken down into subgroups. If a particular subgroup had a high risk, and was a small group, that increased risk would be averaged out by the general statistics. An example would be those women who have a history of spontaneous abortions after the fourth month of pregnancy (when cytological reasons for such spontaneous events would no longer be expected, and obstetrical reasons might be operative). Is their risk from the procedure less than 1 in 1000 or is it higher, and how would one counsel such a woman about the risk of amniocentesis? Should the counsellor use the general risk figure? Should one add that there is no information on whether their risk is the same? The problem of a small atypical subgroup and statistical averaging, or the more general conceptual problem of the single case, are important issues in probability theory.

Case Study 41. Duchenne's Muscular Dystrophy

A young couple with a family history of Duchenne's Muscular Dystrophy (DMD) (usually fatal to afflicted males by age 20) had amniocentesis done to determine whether the fetus was male. The wife's two brothers had DMD and she has a 50% chance of being a carrier based on the mother's carrier status. Muscle enzyme tests (CPK) resulted in normal diagnosis for her; she had three done. The CPK test is 85% accurate at best, and therefore there remains at least a 15% chance that she is still a carrier. The risk figure given to her was 1 in 24 of her child having DMD. Amniocentesis indicated a male fetus, and a saline abortion was performed. Did her brothers' illness influence her risk perception, since she intimately knew the effects of the disease? Was her perception of risk overly pessimistic?

Case Study 42. Spina Bifida

A 30-year-old, pregnant woman had a previous spina bifida child who is hydrocephalic, incontinent, needs leg braces, and has serious learning difficulties. Another pregnancy miscarried. The latest pregnancy is by a different father. At the time, the alpha-fetoprotein test was experimental. There was a 10% risk of not diagnosing a spina bifida fetus and ultrasound was still not sufficiently sophisticated to aid diagnosis. The woman said she would not think twice about aborting, and the decision was less difficult for her than for parents with a less seriously affected spina bifida child. Will a 10% risk be too great for her? In a genetic disease that varies considerably in severity, how should the risk figure be evaluated?

Case Study 43. Cleft Lip/Palate

The couple's only child had a cleft lip and palate, which was the reason they came for counseling. The wife has reservations about having another child. A cousin has had cleft lip and palate repair. Their child has some speech impairment, but the parents view the defect as extremely serious (they would not want to pass the defect on, for example). The physician eliminated a dominant or X-linked mechanism. Although a recessive is possible, the more probable mechanism is multifactorial, with a 1 in 20 recurrence risk. How will this figure be interpreted by the parents in view of their stated determination not to pass this condition on? The risk figures take on varied meanings depending on the evaluation of the problem. Is the application of the figures (seeing them in a context of human interests and purposes) an integral part of their meaning, or is there an objective meaning separate from this context?

Case Study 44. Meckel's Syndrome

A couple had been counseled because of advanced maternal age, were presented the risk figures for Down's syndrome, and on the basis of those figures decided against having amniocentesis. In evaluating Down's, alphafetoprotein and ultrasound tests are routinely done. Such procedures would probably also have picked up a case of Meckel's syndrome. The physician geneticist, a few months later, received an early morning call from the obstetrician who had just attended delivery of this couple's infant; it had Meckel's syndrome and died at birth.

Risk figures presented for a specific problem are not the only risks involved in a pregnancy. With lack of any positive family history for a problem, or any indication that something might be wrong, prenatal detecion programs cannot eliminate all possible risks, and to view them as guaranteeing a perfect baby would be a great mistake. Did this couple make a reasonable risk-taking choice, or were they excessively risk-taking? How do we distinguish normal risk-taking behavior from self-destructive behavior?

Case Study 45. Inconsistent Risk Figures

This case was previously discussed under Denial and Reality Testing (Case #29), as illustrating the possible hostility and resistance raised when the denial defense is threatened. There is a second part to the story. The couple called back to have the test, with the husband now very positive he wanted it done. The full story centered on risk figures. They had previously been told (or had inaccurately heard) that the recurrence risk was 1 in 2000. When told by the counselor that the figure was 1 in 20, they had been in shock for the remainder of the session, trying to handle the changed perception of risk (and it was quite a change). A recurring problem in genetic counseling is the receiving of varying risk figures from different sources. The same situation occurred with the Meckel's case above. The obstetrician told them after the birth of the child that the risk was 1 in 20. Their risk is actually 1 in 4 in each pregnancy. They elected to try again, with the aid of the tests, and again a Meckel's syndrome-afflicted fetus was diagnosed.

The existence of prenatal tests can also affect perception of the risk figures. It is less risk-taking in terms of potential consequences if there exists a test that can detect the presence of a genetic disease; in that event the increased number of options for the patients also effects the meaning of the quantified risk.

A similar case of misinformation involved counseling a couple about porphyria. The husband, who had porphyria, had been told by a family physician that the chance of passing on this condition to his children was one in a million. He was not prepared for the actual risk, 1 in 2. He handled this new risk figure by saying a fifty/fifty chance is about all life ever gives us. However, he was appreciably sadder. His wife was already pregnant when the realistic information was given him, and there is no prenatal test for porphyria.

This patient's attempt to relate the high risk figure to what he perceived as an equally high risk in daily living interprets the risk figure in a distorted context since his view of everyday risks is unrealistically high. This is understandable. In his case, the dice have already been thrown, a risky gamble has already been taken, and his risk-taking decision was pre-empted by false risk figures. Moreover, he has distorted the context of those figues in order to make their meaning seem less risky to him.

Case Study 46. Myelomeningocele

The couple's first child has a rather mild expression of spina bifida, standing and walking, although bowel/bladder control is not certain. A shunt to prevent hydrocephalus was also necessary. Abortion is not an option for this couple and what they really want to know is whether the counselor can tell them, pre-conceptually, whether a child will be normal. Only a recurrence risk figure (3–5%), not a guarantee, is possible. They perceive a 3–5% risk as high, and may choose adoption rather than further pregnancies, even though their first child is doing fairly well. They have contact with a Birth Defects Clinic and are aware of the range of severity for spina bifida.

Has this influenced their perception of the risk figure, putting it in a larger context? Should this sort of reality testing be encouraged for everyone having to make reproductive decisions? Should the negative results of risk-taking be experienced directly, or would the impact be so overwhelming as to seriously distort the meaning of the risk figures. This kind of reality testing implies that: (1) the physician knows the patient's psychological needs and interpretations of experience, (2) is willing to be interveningly directive, and (3) is willing to breach the patient's denial defenses.

Case Study 47. Dilantin Maintenance

A 30-year-old pregnant woman has been taking dilantin since she was three and must continue its use. Dilantin has been tentatively implicated in some birth defects: cleft palate, heart malformations, retarded growth, and finger malformations. There is an estimated 10% risk. The physician deliberately worded this in a positive sense: "There is a 90% chance of having a normal baby." Both statements are accurate. Is their meaning, however, the same? The physician can influence, sometimes strongly, how risk figures will be perceived. If concern about raising anxiety is a great concern, the physician may emphasize the positive aspects of the risk figures. If the possibilities for medical intervention are limited, the physician may also emphasize the positive aspects because of not wishing to create unnecessary (unproductive) anxiety.

Case Study 48. Amniocentesis with Questionable Indications

A young couple, both lawyers, want only one child and want that child to be healthy. They were interested in the possibility of multiple testing. Although this could theoretically be done, the impracticality of it, in the absence of positive family history or other indication of heightened risk, was stressed. Since he is Jewish, a Tay-Sachs carrier test was done at his request. Their risk, however, is very low. They wanted amniocentesis as well, and the physician agreed to do the test. The resident questioned whether amniocentesis was reasonable in this low-risk situation for any defect. The physician left the choice of this to the fully informed couple, since his position is that the doctor should not directively decide what is a meaningful risk. As this husband had said: "It's *my* risk." However, if the procedure involved a greater risk than amniocentesis (fetoscopy, for example), the physician would become more negatively directive, and at some point would refuse to do the procedure personally.

This case illustrates the possibilities of conflict because of different risk perceptions. It is true that, as the husband said, it is the couple's risk and that the meanings derived from the risk figures must be their own. Their interests and feelings are central to them and they interact with a physician in order to advance those interests and feelings. The physician accepts this as part of the definition of the genetic counselor's medical role, as one of the primary parts of the definition. But it is not the sole element of that role. The physician is not a hired instrument. Rather, another part of his or her role definition is that of expert, and part of that expert knowledge does involve evaluating the risk/benefit ratios of medical procedures.[32] The very role implies that activity. The physician can force the patient to assume what may be considered a minor risk if that physician believes the risk of the medical procedure to be a major one, and in truth we would want precisely that done. Otherwise, a physician would be assisting in the self-destructive behavior of the patient (as the doctor sees it). If most professionals also view that behavior as self-destructive, the decision is reinforced. The final question, of course, is whether social role-playing should block or allow self-destructive behavior.

The position whose brief structure I will give here differs fundamentally from the usual justification in medical ethics for intervention. It is a general argument for any intervention at all, although I am phrasing it in terms of self-destructive behavior. The three usual philosophic bases in the literature are: (1) respect for persons (the Kantian argument), (2) the utilitarian calculus argument, and (3) the Rights Model.[33] The respect-for-persons basis founders on a search for its own suitable justification. Kant's attempt to ground it on noncontradiction is not satisfactory since we can always question our acceptance of noncontradiction (as Freud pointed out, the human mental process normally accepts and exhibits contradiction). Its exclusively rational characterization or emphasis cannot carry the burden for intervention and "respect" usually ends up being interpreted in terms of *norms* for human well-functioning; thus we come back to the need for an actual systems approach to establish the real foundation for medical intervention. The utilitarian calculus argument, although institutionalized in the form of bureaucratic cost/benefit analyses,

    [32]David L. Bazelon, "Risk and Responsibility," *Science* **205**, 227 (1979).
    [33]For an example, see B. Eichelman and J. D. Barchas, "Ethical Aspects of Psychiatry," *Psychiatry and Behavioral Sciences*, D. A. Hamburg, J. D. Barchas, P. A. Berger, and G. R. Elliott, eds., Oxford, New York, in press.

has serious problems with comparing the various degrees of happiness, with the mixing of social and individual levels of analysis in its hedonism, and with its failure to come to grips with the logic of Silenus' calculation that it is probably less painful never to have been born at all. The "rights" basis for medical intervention is in desperate need of a basis for "rights" itself, especially if we do not wish to assume a supernatural lawgiver or an unchanging natural order. Attempts to ground medical intervention in social rights run aground either on the problem that what society gives, society can also take away; or on the problem that, even in a hypothetical social contract or original position situation, actual human behavior is frequently risk-taking, all-or-nothing gambling, noncompromising and passionately committed, and self-destructive—in effect, the full range of our nonrational possibilities. I believe the justification I here propose avoids all of these difficulties.

I can only sketch an argument here:

1. There is no philosophic reason to prefer self-destructive desires to self-preserving ones. Such affects are primary and we need to assume self-preservation or well-functioning to generate the cognitive values that philosophizing is based on.

2. However, the decision on how to respond to self-destructive behavior is usually made by people within a functioning, preserving context. Their shared interests and goals, based on the self-preserving assumption, can conflict with self-destructive behavior. If the conflict is potentially damaging to them, concern and compassion for the autonomy of the self-destructive individual may be outbalanced by considerations of the damage to their own interests and goals.

3. Social roles incorporate this calculus, which should, of course, always be reasonably and accurately done. A terminally ill person who wishes to commit suicide has only a limited capacity to damage society or the circumscribed circle of personal intimates, and compassion and concern for the potential suicide's needs should not be overridden by unrealistic fears. Occasionally, the distress of someone engaged in self-destructive behavior may be so great that calculation of damage to others may be superceded.

4. Social roles tend to preserve calculations of consequences that were once realistic, but in view of changing environments, are becoming inadequate. Therefore, role responses need to be adapted to feedback. If enough patients insist on taking a particular risk, or if one patient very strongly

argues for personal acceptance of a certain risk, the physician's responses in terms of traditional or previously learned responses should then be seriously reconsidered. Filling roles must not involve being locked into patterned responses.

5. The social process, through sanctioned roles and consensus, attempts to distinguish between normal or healthy risk-taking and self-destructive behavior. This is not an easy distinction conceptually to draw, at least when it is restricted to behavioral considerations, or when the avowed purposes are suspect. We usually sanction behavior such as Grand Prix racing as very highly risk-taking, but not as pathologically self-destructive. We begin to suspect that some teenage males who wrap their cars around trees with reckless abandon are death-seeking, although perhaps ambivalent. Although we routinely list the cause of death as accidental, we believe, without being able to prove it, that a middle-aged businessman under some severe life stress, whose car slams into a bridge abutment, actually committed suicide.

How do we distinguish between the three? When is it risk-taking and when is it self-destruction? Purpose or motive, self-reported or inferred, makes a large difference. The racing driver presumably wants the fame, fortune, and excitement involved in the risk-taking, and not death. The driver's training and expertise involve self-preserving skills, which is a behavioral indication that corroborates his or her stated purpose. The institutionalizing of the risk also expands it or externalizes it, so that the driver's private interests are not the sole factor. The businessman, on the other hand, has an inferred interest in ending his life and little purpose in reckless driving. The behavior exhibits no self-preserving components, and private interest in driving under unsafe conditions fits into no grander scheme. He has not died for a cause or a goal beyond an internal one. Thus, although accidents *do* happen, this risk-taking seems better described as self-destruction. The free-wheeling teenager, however, is more difficult to place into this scheme of distinctions, and combines elements of both. Self-preserving behavior is present in his pride of driving skill: how well the machine can be handled, how well spatial conditions are judged. The teen's purposes are partly externalized as an element of a young driving culture bent on exhibiting and admiring those skills. On the other hand, adolescent stresses may be strong enough to generate some darker motives. Furthermore, the social judgment may depend on how this age group is perceived. One society's rite of manhood may be another's suicidal impulse.

When the genetic counselor analyzes the risk-taking decisions of patients, the same conceptual problem arises. In addition, there are the added consequences to a defective child whose parents have lost their genetic gamble. The risk is an extended risk. There are two choices made here, one involving the value we place on autonomy (it is *my* risk) and the other involving the value of intervening to prevent pain and suffering that may be more extensive than the parent's distress alone (it's the child's risk too). If risk-taking is viewed as self-destruction in some cases, we have the same ethical problem as that involved in suicide. For a social individual, consequences extend beyond the immediate personal ones, as Aristotle perceived, and as the second part of my argument states. Not only social macro-ethics is involved here, however. On the micro-ethical, personal-interaction level, other individuals also have an interest in not living in a world of unnecessary pain. We empathize, and that empathy makes us hostage not only to history, but to those persons we come in prolonged contact with. Risks are ultimately shared, a factor that needs to be included in the ethical equation. It is important to make a reasonable evaluation of risk-taking or self-destructive behavior because that activity does involve us potentially. If the decision is made that an act is self-destructive, we, as the individuals involved, have a vested interest, as well; such an act will challenge our own well-functioning, and for that reason, we are ethically justified in considering, and perhaps undertaking, some intervention.

There is some literature on conceptual problems generated by concern for the "possible" child who might be born defective (Baker et al. [33]). How can "possible individuals" have rights to happiness, or be owed utilitarian debts by us? Actually, I prefer not to use the term "rights," a position I share with Richard Taylor.[34] Rather, how can we consider the interests of "possible individuals"? We can only consider them as our *projects*, to speak in an Existential sense. It is our interests, our purposes that are at stake, because we can project a future of unnecessary pain and that future is realistically based on previous experience with the pain of actual children. This is a project we can and do choose against, but its ethical justification lies in ourselves, not in the "possible individuals." It is our project, our affirmative attitude,

[33]Robert Baker, "Protecting the Unconceived," *Contemporary Issues in Biomedical Ethics*, Davis, Hoffmaster, and Shorten, eds., Humana, Clifton, New Jersey, 1978, pp. 89–100; Michael Bayles, "Harm to the Unconceived," *Philosophy and Public Affairs* **5**, 292 (1976).

[34]Richard Taylor, *op. cit.*

and our perception of pain that forms the choice. This too is a required component of the ethical analysis. Just as the physician does not have to collaborate in a patient's self-destruction, so too we as individuals do not have to collaborate, nor does the social system have to collaborate. The trick is to make sure the attribution of "self-destructive" to a behavior is an accurate label.

Throughout these arguments, I have been assuming that the risk figures can best be interpreted as gambling behavior, which presumes the solution of some specialized issues in probability theory. There is a subjective–objective battleground here, imbedded in the frequency, personalist, and propensity theories.[35] Frequency theory, though assuming a correspondence to reality, has a problem in the handling of single cases and in assigning them to the appropriate reference class. This is critical to genetic counseling because genetic counseling uses frequency risk figures (figures based on an entire class), but assigns these figures to individuals. Personalist theory, while emphasizing betting behavior, risk-taking, and expert's decision-making, may abandon correspondence with reality for a dubious coherence in a single case context. Propensity theory gives us rather esoteric chance and disposition constructs as subjective–objective proposed solutions.

The actual components of the case studies seem to favor risk-taking or gambling behavior, with two connections to more than a subjective interpretation: (1) the reality testing of direct experience with the genetic problem, so that one can project what loss of the bet really involves; and (2) the risk figures that are variously interpreted but do give, for the long-run, the frequency of losing or winning the gamble. A study by Barnert, Grobstein, and Miller at Stanford illustrates all these components.[36] A burden instrument (questionnaire) measured the subjects' reactions to the burden of simply described, nonlabeled genetic conditions. The risk the subjects were willing to assume for each condition was measured. The higher the perceived burden, the less risk subjects were willing to take. The perceived burden, of course, is dependent on life experiences and information. One group, a random, nonpatient group, showed higher perceived burden when the instruments were first administered. After counseling, their perceptions of risks had

[35]D. H. Mellor, *The Matter of Chance*, Cambridge University Press, Cambridge, 1971; Henry E. Kyburg, Jr., "Chance," *Journal of Philosophical Logic* **5**, 1976; Wesley Salmon, *The Foundations of Scientific Inference*, University of Pittsburg Press, Pittsburg, PA, 1967.

[36]C. R. Barnett, R. Grobstein, and W. B. Miller, "Measuring Reactions to Burden and Risk with Standard Instruments," *Excerpta Medica*, International Congress Series, No. 397.

been lowered. The counseling added some reality testing to the gambling behavior, which significantly influenced the projected behavior. Risk-taking, when reality testing is added to the process, seems an adequate description of how real individuals handle genetic probability figures, but this empirical-level solution will probably not adequately handle the theoretical difficulties with probability theory itself.

Here, I would like to add one last comment about risk itself. Frequency rates for genetic risk put the individual problem in a broader context, but this context, to make sense for life decisions, needs to be even further broadened. As Bazelon says,[37] few favor risk for risk's sake. Yet risk, like anxiety, is an integral part of living. He points out that to have energy, we incur risks from energy production. To have mobility, we risk accidents and air pollution. To have a cheap food supply, we risk carcinogenic additives. There is no choice between having or not having risk at all, but rather a choice of how much, from where, and by whose decision—tragic choices often. In making those value choices, reality testing in terms of a broad context is helpful. We should perhaps all carry around Wilson's risk chart,[38] which lists common actions that result in a one in a million chance of death: smoking 1.4 cigarettes, or driving 2 days in New York or Boston, traveling 10 miles by bicycle, traveling 300 miles by car, flying 1000 miles by jet, having one chest X-ray, or drinking chlorinated water for 1 year, for example. In some cases, we have no choice but to indulge in the activity. What prenatal detection does in many cases is increase the number and power of our risk-taking choices, expanding our options in the face of always present risk.

Physicians may sometimes misperceive risks and interpret a patient's problem as representing much more risk than is actually involved. Such misperceptions lead to overmedicalization.

Overmedicalization or Good Medical Practice?

Case Study 49. *Achondroplasia*

A couple was counseled for possible achondroplasia, one of the types of dwarfism. The wife is 4'10", with short limbs. Her father was 5'1" with a long trunk and short limbs. She had not been concerned about her body form, is in good health, and is a physical

[37]David L. Bazelon, *op. cit.*

[38]Richard Wilson, "Comparing Risks," *Rochester Democrat and Chronicle*, June 10, 1979.

therapist. Her gynecologist had given her information on achondroplasia when she told him she wanted to start a family.

Is this overmedicalization? Would she have been concerned unless her physician had been? Her condition does not appear to be classic achondroplasia. If her doctor had discussed the matter with a consellor first, perhaps the physician would have been sufficiently reassured to avoid involving the patient.

Case Study 50. Suspected Down's Syndrome

This is a 13-month-old baby being considered for adoption. His young natural parents were on drugs during the pregnancy. There is a simian crease (I know a neuroscientist with simian crease), possible epicanthal fold, and low-set ears. Development has been completely normal. Physical examination indicates a probably normal child. A chromosome analysis was done.

The adoptive parents' physician had made the referral, with the parents knowing little of any possible problems. The referral and the medical procedure, however, might well have an effect on how the prospective parents will view this child. Was the physician exaggerating the risk? Would that doctor have perceived as great a risk if the baby were the couple's natural child? Does the doctor have a lower risk-taking level for adopted children and should adoption imply a greater guarantee of health (less risk) than a natural birth?

Case Study 51. Schizophrenia

This unmarried young woman is 20 weeks pregnant. Her obstetrician requested the counseling because it was felt that a therapeutic abortion should be considered. She had been a psychiatric inpatient and was given medication, although she had suggested to the staff that she might be pregnant. The physician is concerned about the medications, her competency as a mother, and the genetics of schizophrenia. She discharged herself from the hospital. During the post-discharge counseling session, her primary concern was to find out what her problem had been diagnosed as. The degree of fetal risk from her use of psychoactive drugs is unknown.

Was her doctor being unduly concerned about her risks? Should the obstetrician have intervened in the problem of her competency as a mother, or restricted the terms of the discussion to the medication and genetics problems? Should the genetic counselor have handled only the medication and genetic transmission risk?

In my discussion of the unfragmented patient and the proper role of medicine in Chapter 1, I believe those questions have been answered. What the genetic counselor actually did parallels the recommendations that emerged from that discussion. With the assistance of a psychiatrist, the physician answered her questions about diagnosis, reassured her about the risks resulting from medication, and using the leverage of a good interaction, influenced her to seek continuing therapeutic support.

Overmedicalization is an evaluative level that depends heavily on reality-testing and on affective responses to the amount of pain in the world and the prospects of relieving it. The latter is modifiable to some extent, but is hardly under the rational control we would like to think. A physician who continues, however, to ignore the negative feedback that results from his or her hyperperception of risk levels and the costs of intervention would be incompetent. That sort of sensitivity and attention to the environment is what we mean by the "art" of medicine. Far from art, it is actually good empiricism in a pragmatic sense: responding to experience, and adjusting or modifying the system as a result of feedback loops.

Summary

These initial three chapters should have indicated that philosophic analysis, as practiced in the context of applied medical ethics, needs to expand its base beyond narrowly rational considerations. The alternative is not unreason, since the process of reality testing and the latter two clauses of my hypothetical ranking rule must be reasonable processes. Reason remains an important part, but the analysis is incomplete if it stops at that point. Valuing, choosing, living, or experiencing is much more complex than that. Reality testing is done in a context of affective responses to existence, to pain, and to pleasure. It involves purposes, interests, and needs that do not fit the rational model. The first clause of the hypothetical ranking rule concerns attitudes toward the universe, primary motivating emotions that can later be characterized as well-functioning or malfunctioning, self-preserving or self-destructive, by those with an affirmative attitude, but from a theoretical neutral perspective simply *are*. Ethics needs to be extended to include the nonrational (I emphasize again, not the *irrational*) in order fully to describe an axiological system. To leave the nonrational element out of such analyses is to create distortion and a plethora of unnecessary conceptual problems. Empirical psychology, not philosophical psychology, can be of considerable assistance in this task. The justification of such a program is beyond the task of this book, and I merely hope successfully to have outlined its rationale. But as the case studies illustrate, the need for such a more complete analysis is pressing.

The Systems model used here also supplies an effective analysis of instrumental/intrinsic good concepts, making such distinctions a matter of range of purpose or breadth of interest. Systems analysis also points to the need for a balance between macro-ethics and micro-ethics. In addition, it points out the adaptive, changing nature of our decisions.

The social context of any ethical decision-making process is an important consideration. Within this context of social bonding (and therefore, empathy), it has been argued that our choices are made in terms of our projects, an extension of our needs and interests to anticipated needs and interests. The justification of intervention, and a clearer analysis of self-destructive behavior, were offered within this framework.

Chapter 4

Self-Image

The Worth of the Individual

The worth of an individual is one of the fundamental ethical issues. How do we value human life in general? How do we value (and sometimes choose between) concrete individual lives? Is it always wrong to choose against human life, or do we implicitly make that choice daily and only shy away from explicitly stating it? Is human life special in the universe?

In this chapter, I want to look at the concept of "worth" and will attempt to determine what factors lead one to a positive valuation of an individual. If I can identify these touchstones clearly enough, they will also help clarify how individuals evaluate themselves. There are two valuings involved, then: external (made by intimates, subgroups, and culture) and internal (made by oneself, granting that this is understood to be the internalization of learned external valuation, colored by the individual's unique experiences).

Let me in the following sections briefly enumerate some of the components of self-image.

External

(1) Each human life is intrinsically worthwhile. This is often a stated general cultural principle. Whether we behave as if we accept it is another matter. Is it an empty principle? What are the grounds for holding it? Does simple existence equal worth?

(2) Well-functioning human lives are worthwhile. "Well-functioning" can be defined in a number of ways: successful (famous, wealthy, powerful), happy (ecstatic, content, accepting, self-understanding, fulfilled, sensuous), healthy (performing, intact, free of disease, minium level of experiencing), contributing (important, extending possibilities).

(3) Potentially well-functioning human lives are worthwhile.

(4) Subgroup-model human lives are worthwhile. Besides the various evaluations of cultural heroes (a consensus approach), there are subculture heroes as well.

(5) Subgroup potentially well-functioning human lives are worthwhile. The well-functioning in this case is evaluated within the subculture, not the consensus, context.

(6) Human lives emotionally important to the individual concerned are worthwhile. For whatever reason—habit, meeting of present needs, unconscious wishes, physical attraction, shared experiences, kinship systems—we value some individuals in a special way, viewing their worth as crucially central to our own existence.

Internal

(1) Each individual, simply in view of experienced personal existence, views her- or himself as worthwhile. This generalized self-acceptance and self-interest or self-love, contrary to currently popular psychological analyses (e.g., that of Christopher Lasch[39]), may not be as empirically demonstrable as the cultural castigators would have us believe. Narcissism is hardly one of the leading diagnoses of mental illness in our time, being easily crowded aside by an array of popularly accepted self-repressions, self-hatreds, guilts, punishments, and that rapidly rising mortality chart category, suicide or self-destruction.

(2) Individuals view themselves as worthwhile because they function well in one or a combination of the following spheres:

(a) Physiological. Individuals are pleased with their bodies and find their functions valuable.

(b) Intellectual. Individuals are happy with their capacities to manipulate symbols in logical and creative ways.

(c) Emotional. Individuals find their emotional responses to be of great worth to themselves and/or to others.

(d) Social. Individuals feel they are contributing in some way to the group, and are important in the processes of the community.

(e) Manipulative. Individuals can make or create something, can do something well.

[39]Christopher Lasch, *The Culture of Narcissism*, Norton, New York, 1978.

(f) Ethical or religious. Individuals have worth because they view themselves as good persons, people who try to do the right thing.

(3) Individuals view themselves as worthwhile because they can potentially function well in one or more of the above areas.

(4) Individuals view themselves as worthwile because someone else does. This can run the range of external groups.

(5) Individuals are ambivalent.

Evaluations of personal worth do not take place in a vacuum and the cultural baggage of centuries often generates responses that conflict with what we would expect to be reasonable in view of current scientific information. In addition, and especially critical for genetic counseling, internal perceptions of any person's body have been developmentally built up from early infancy, and many nonrational factors may have invested the body with unexpected symbolic meanings and importance. When we combine notions of "bad blood = inferior social status and individual worth" with the feeling that the body, once an important source of pleasure, is now a source of betrayal, and add to that the socially ancient religious and valuational mystiques of reproduction, we have a highly volatile situation. There has been concern, therefore, that genetic information is potentially very damaging to an individual's sense of self. In this chapter, I will look at this issue from a number of perspectives: that of the identified carrier of a genetic disease, that of a parent affected by the disease, and that of the offspring affected by it. Again, it will be helpful to begin by looking at some actual case studies.

Carrier Status

Although carriers of a dominant gene are certainly in this group, the cases will be restricted primarily to recessive, X-linked, and polygenic mechanisms. Except for incomplete penetrance dominants, the dominant gene carriers will be discussed under the section dealing with affected parents.

Case Study 52. Tay-Sachs

A young Jewish couple, the wife 2 months pregnant, came for fetal Tay-Sachs testing. They are both carriers for this recessive trait, which they discovered during the Washington mass screening program. This program combined not only the screening element, but education concerning the genetic problem and what options were available for detected carriers.[40] The couple had been shocked to discover they were both carriers since neither of them was aware of any family history of Tay-Sachs. However, the family pedigree was very incomplete; World War II and concentration camps had eliminated all their relatives. They have decided, with the help of the prenatal test, to reproduce. The Jewish community not only actively supports Tay-Sachs screening, but supplies an accepting social environment for carriers. It is supportive in both senses. This may make a crucial difference in the response of identified carriers, since the subgroup is reinforcing an individual sense of worth in an active way.

Case Study 53. Mucopolysaccharidosis, Hunter's

The young woman is 17 weeks pregnant and came in for a prenatal test for Hunter's syndrome. Her brother died at 14, severely mentally retarded, confined to a wheelchair, and suffering respiratory problems. The Hunter's syndrome diagnosis was confirmed by a very reliable recent test. Although autopsy was refused, the tissue culture gave a definitive answer. She and her mother had a less reliable carrier test that indicated both were carriers of this X-linked gene. However, the physician could not say with certainty that she was a carrier, only that it was probable. The timing on this test will be very tight. Cell culture growth will take perhaps 4 weeks, although sex determination will be made sooner. If the fetus is a girl, there will be no problem. However, if it is a male, the lab will have to send the material to another lab specializing in mucopolysaccharidosis testing, and this will bring it dangerously close to the 24-week abortion limit. This was not a planned pregnancy, the woman does not yet have an obstetrician, and there is some question about the delay in contacting the counselor. She had been fully informed, at the time of carrier testing, about the importance of seeking help early in the pregnancy. Is she really nonchalant about her carrier status, or rejecting of the information? Did the uncertainty of the carrier test make a difference?

[40]Michael M. Kaback and John S. O'Brien, "Tay-Sachs: Prototype for Prevention of Genetic Disease," *Hospital Practice*, March, 1973, pp. 107–116.

Case Study 54. Epidermolysis Bullosa and Legg-Perthes Syndrome

Family history on the wife's side involved both epidermolysis bullosa and Legg-Perthes (epiphysial hip dysplasia) syndrome. This form of epidermolysis bullosa is an autosomal recessive characterized by skin blistering and nonhealing lesions that can be lethal or sublethal, and three of her four siblings died at a quite early age. She has two in three chances of being a carrier. Legg-Perthes can be a dominant trait. The couple is not concerned with either problem. Her risk of carrier status for a lethal or sublethal gene has not seemed to affect her self-image in any damaging way.

Case Study 55. Possible Trisomy 18

The couple have a healthy 2-year-old and had just lost a baby with possible trisomy 18, a chomosomal abnormality characterized by mental retardation and heart defects. This is a young couple, so that a balanced translocation mechanism would be suspected. However, no chromosome studies were done on the baby to confirm the diagnosis. The couple wants one more child, but the wife would prefer to wait. If she became pregnant, she would want amniocentesis and chromosome analysis of the fetus—but she does not want a chromosome test on herself. The test would only involve drawing blood, but this is apparently information she does not want. Is she having image problems even considering that she could be a carrier for a genetic translocation?

Case Study 56. Cystic Fibrosis

There is currently no carrier testing or prenatal diagnostic test for cystic fibrosis. However, this couple came for counseling, since the wife is 16-weeks pregnant and concerned about the positive family history. Four of her six siblings died at very early ages from CF (the youngest in 7 weeks, another at 1 year). Her risk of being a carrier is 2 in 3, but her husband's risk is 1 in 25. She was open with her husband about this CF family history, but very disapproving of her brother who had not told his wife about the problem before their marriage. Her brother's attitude toward probable carrier status was quite different from hers. Why had he chosen not to reveal this genetic information about himself? Did this genetic aspect of himself figure importantly in how he viewed himself, how he valued himself? Was he afraid that others would change their evaluation of his worth? Did he feel a possible reproductive problem would seriously devalue him in the eyes of his fiance?

Case Study 57. Cystinosis

This family has five of the six children affected by cystinosis, an autosomal recessive involving the accumulation of cystine crystals in major organs (brain, kidneys, eyes). Two of the children died early, one with renal failure, one with kidney transplant failure. A boy is currently being treated symptomatically, and two others are presently asymptomatic, but tested and diagnosed as having the disease. There is only one child who does not have cystinosis. Both parents are carriers. However, this knowledge has not seemed to affect their reproductive plans or their perception of worth. Should it have changed their attituude toward reproduction? If they do not view the genetic defect as influencing their worth or their children's worth, have they the right to continue to reproduce? Does a sense of worth imply no ethical basis for objections to carriers for a genetic disease having children who will be very seriously affected by that disease? Or are the two issues separate and not to be confused?

Case Study 58. Spontaneous Abortions

The wife had just miscarried a third time and has a maternal family history of numerous spontaneous abortions. She's very discouraged. The husband's history is negative; everyone seems to be very healthy. Chromosome testing was done to rule out the possibility of a balanced translocation. The test can eliminate the possibility of an observable chromosomal genetic problem, but not the possible presence of a lethal recessive. At this point, the wife feels the fetal deaths are her fault. How does the concept of "fault" fit into a genetic description of the problem? If something is physiologically abnormal about a person, how does this translate into personal fault and should it?

Case Study 59. Thalassemia Screening

This area has a routine multiphasic screening program at health centers. One of the tests run on all patients is a thalassemia carrier test. If the results are positive, the patient is asked to come in and a physician then offers information about and explains the problem. A young woman about to be married was identified as a carrier. She then asked her fiancé to check his carrier status, which he reluctantly did. He learned that he was not a carrier for thalassemia, but rather, had sickle cell trait. The physician could not convince him to discuss this with the girl. Instead, he and his mother decided it would spoil the wedding. Either his image, or how he feels his fiancé will perceive his worth, or both, are threatened by this genetic information. In addition, the defective gene he carries is more usually associated with another racial group. For certain genetic problems, carriers can vary from the norm, sometimes enough to cause minor physical problems, but often not enough to show any clinical signs at all. Should the presence of an abnormal gene be causing the shame it apparently is? Why, in some of these cases, are people making this so important a factor in their sense of worth or of self as they are?

The damage we do to our sense of worth, we do ourselves. But since we are a social species, and since there is a time lag before new information is handled properly by a system, we often have a great deal of help in this self-derogating process. Identification of carrier status, like the unexpected birth of a defective child, is especially disruptive because, unlike the cases of affected parents who already know about the problem and have to live with it, it is a sudden and surprising trauma. Suddenly we explicitly leave the ranks of the "normal," even if we have never clearly understood what that meant in the first place. These experiences occur in a cultural setting that has primed us to consider reproduction a supreme value and pleasure, and that therefore reinforces a self-devaluation if there are reasonable considerations for not reproducing or for needing considerable medical intervention to reproduce. Instead of one of many components of our person presenting us with difficulties, we are programmed to believe that our entire existential self has lost worth, and indeed there is little cultural support to correct that perception. The status of the products of our gonads and our reproductive behavior thereby assumes an importance much out of proportion to the quality of our actual and potential living. We begin to define ourselves predominantly in reproductive terms. Although I disagree with the Sartrean existentialists that there is no objective human nature, my views are certainly compatible with John Wild's[41] suggested pragmatic answer to the object–subject problem: Is there a human na-

---

[41]John Wild, "Authentic Existence: a New Approach to 'Value Theory,' " in *An Invitation to Phenomenology,* James M. Edie, ed., Quadrangle Books, Chicago, 1965, pp. 59–77. See also, Asher Moore, "Existentialism and the Tradition," *An Invitation to Phenomenology,* pp. 91–109; Jean-Marie Benoist, "Classicism Revisted: Human Nature and Structure in Levi-Strauss and Chomsky," *The Limits of Human Nature,* Jonathan Benthall, ed., Dutton, New York, 1974, pp. 20–48.

ture separate from our making it? I believe we can safely use an expression such as "defining ourselves" if we mean by that picking human potentials we wish to emphasize or actualize, and making that selection on the basis of past experience, all within the context of our limiting 'objective' natures (a nature that can be adequately generalized for all of us but not externalized, apart from our psychological activities). We make ourselves, but we do it within the reality of a world we are part of, not separate from.

Thus our definition of ourselves as primarily reproductive, embedded in our cultural tradition (especially in the humanities) and usually individually internalized, is the first theoretical obstacle to disentangling our sense of worth from reproductive capacity. Screening, ethical concerns about creating pain, social costs, dysgenics, therapeutic or selective abortion, population concerns—all are seriously affected by the cultural tie between worth and reproduction. The old economic base for this is shattering, but any change, because of the time delay built into any complex feedback system, may be painfully slow.

Part of a genetic counselor's task is to swim against this cultural current and help people identify other equally worthwhile and important aspects of themselves. However, unlike the situation in the Tay-Sachs case, where subgroup support was available, there is little general cultural support for identified carriers of a genetic problem. Even there, I could have chosen another case where a couple, having lost their baby girl to Tay-Sachs, used our prenatal testing and subsequently had two healthy children. However, they did not want blood tests done on the children to determine whether they were homozygous or heterozygous (carriers), and even here, possible carrier identification seemed threatening to them.

To compound things, we also have a cultural and theological history that associates our sexuality with reproduction: the purpose of sex is procreative. As a result, all the emotional content of our early sexual experiences and our sexual image of ourselves becomes connected to reproduction. Sterility becomes associated in our minds with lack of masculinity or femininity. Not having healthy children becomes a devaluation of our sexual selves: we have failed as men or women, and a sexual definition is a rather primary one to feel a sense of failure about. To be informed that our DNA code ranges from mildly garbled to lethally unintelligible then also becomes equated with our sexual adequacy by this tortuous conceptual route.

The assumption at fault, of course, is that sexuality = procreation. The equating of worth with normal reproductive capacity can therefore be seen to involve two major, empirically unsupported assumptions: (1) that we are most importantly and primarily reproductive individuals, and (2) that our sexuality is defined in terms of our reproductive ability. If at this point we add the element of personal fault or guilt, then we have created the damaging and unreasonable cultural context in which genetic information is imparted. It continues to amaze me that the recommended solution often is to censor the information available and restrict the gathering of new information, as if the bearer of the news were responsible for the situation.

How fault becomes connected to a physiological abnormality is not apparent. We have a tendency to think complacently that by medicalizing certain abnormal mental processes and social aberrations we have freed people from unwarranted judgments of sin. But we have not been as successful as we would like to believe, even on the level of physical abnormalities. Quite reasonably, no one need initially bear any personal responsibility for varying from the normal range of the human condition, especially physiologically. For example, there are some HLA markers for genetic structures that indicate a predisposition (a susceptibility) to emphysema and lung cancer.[42] To model this data in terms of personal responsibility, free will, fault (can one be at fault unless one is responsible and free?) is archaic and makes no sense. If, knowing this susceptibility, the individual becomes a chain smoker, are we now talking about responsibility, free choice, and therefore fault? In a highly restricted sense, yes, but in the older theological sense of uncaused and morally punishable action, I still doubt it. There are previous life experiences over which the individual had not that much personal control that led to chain smoking. The individual now has the responsibility to accept the unpleasant consequences, although I pull back from calling this fair; it is simply what we mean by being individualized: we accept responsibility for the consequences of our caused actions. If the individual has an affirmative attitude toward existence, then he or she will bear the responsibility of responding to feedback from the environment and attempt to adjust personal behavior to well-functioning. Nowhere in this concept of responsibility, however, is allowance made for the older concept of punishment for being morally bad (not being of worth). West has characterized our culture as embod-

[42]Patricia Cox, "Protease Inhibition," Lecture, Feb. 1, 1979, discussion of Alpha-1-Antitrypsin, elastic tissue destruction, MZ's risk 2–3 times normal risk for obstructive lung disease, Genetics Regulation Series, University of Rochester Medical School.

ying the concept of punishment as a deterrent,[43] and we may now be able to see some of the human cost of such a deterrent. The element of fault, then, enters our equation by way of traditional humanist concepts of free will and moral responsibility, of uncaused actions and choices made from an essentially evil (not of worth) nature. Regrettably, we have not expunged these notions from our attitudes toward those exhibiting physiological variance from a norm, at least not in genetics and reproduction, and I suspect not so successfully either in those diseases where an outside infectious agent is not readily identifiable. As a result, identified carrier status can be a serious blow to self-worth, judged both externally and internally.

There are internal sources for the threat to worth as well, although I have emphasized the social or cultural ones. A sense of betrayal by what had been one's primary means of gratification is important. If one primarily identifies self with body rather than behavior, then the body's worth is threatened whenever it is perceived as varying too far from the norm, and by corollary one's self-image is threatened. These internal yardsticks of self-worth are very difficult, nonrational components to deal with, and thus there will always be some necessary danger to self-image in the conveying of genetic information. My goal is to sketch the sources of the *unnecessary* dangers. It is also important to remember that non-informing cannot be separated from its consequences, and that non-informing would therefore be a choice to allow present processes to continue and to have their effect.

Thus far I have suggested that personal worth and carrier status need to be disengaged. If carrier status says little about the person's worth, if they can indeed be disengaged, does this then remove any basis for an individual's being held responsible not to reproduce? I want to address this question in terms of all the possible clinical situations, so let me first add the context of case studies involving affected parents.

Affected Parents

Case Study 60. Porphyria

This case was previously presented under the category of risk perception. Let me fill it out a bit more here. The affected husband keeps out of the sun as much as possible. When his problem first

[43]L. Jolyon West, University of Rochester Department of Psychiatry Conference on Violence, Rochester, N.Y., June 1–2, 1979.

surfaced, he had extensive face and hand lesions, and there is now some scarring on his hands. He works outdoors and is too active in his inclinations to be content to remain indoors. He uses sunscreening lotion and wears a hat or gloves. His general health is good. He was startled to learn that the risk was 1 in 2 of passing his condition on to his children. The physician asked him how much of a problem his disease had been for him, and he indicated that he managed reasonably well. There are some possibly serious complications that he fortunately has not experienced. The couple wants a family, but in spite of this, and his indicating that he felt all right about his health status, he remained very concerned about passing porphyria on to his children. Although his perception of his own physical state is positive, or at least accepting, he may not wish to be responsible for his children's experiencing the same difficulties. His sense of self-worth does not preclude his judging his genetic condition to be undesirable. This is an empirical indication of an interesting separation between an individual's perception of personal worth and the devaluation of genetic disease.

Case Study 61. Neurofibromatosis

Neurofibromatosis is a dominant genetic trait that needs careful, long-term monitoring. A 26-year-old woman wanted counseling prior to marriage. She and two sisters have the disease, as does her mother. Her married sister has an 18-month-old son with the problem and her 3-year-old daughter just died from it (an optic glioma). The counselor suggested that how this woman feels about herself should determine whether she chooses to have children. This directly links self-image with reproductive choice, making a negative decision to have children also a negative evaluation of oneself. Is this link conceptually sound? Or did the counselor really mean that she should consider how much of a problem this disease had been in living her own life (not that it subtracted from her worth) and imagine her children sharing this same problem? Phrased in this way, the counselor creates no link between self-worth and procreative choices, which is a crucial difference.

Case Study 62. Osteogenesis Imperfecta

Prepregnancy counseling was requested because the husband has osteogenesis imperfecta, a dominant genetic trait with a wide range of severity. His mother also has the disease, but neither of them appear much affected by it. He is healthy, exhibiting the grayish sclera that are consistently expressed with this disorder, and his only problem seems to be a slight hearing disorder. There is a very strong reproductive desire, but there is no developed prenatal test to offer them. The severity of expression of the disease may resemble the mild familial history, or be much more severe, but this is not really predictable. At the moment, self-image, because of the mildness of the disease, is not much of a problem. How would the father perceive his responsibility if a child were born with very serious expression of the disease? When self-worth and reproductive choice are linked, what are the consequences if a very seriously affected child is born?

Case Study 63. Ectrodactyly

An unmarried pregnant woman in her early thirties has ectrodactyly, also called "lobster claw deformity." She had been abandoned by her parents at birth and spent most of her early years in a hospital. One hand has been partially surgically corrected and her feet are affected (when both hands and feet are involved, it is an autosomal dominant). Her father had the condition, as did an aunt. She is employed, but frequently had to hide her hands during job interviews and she is very concerned about the effects of similar negative social reactions on her child, very concerned that a child of hers would have to experience the same social burdens she did. She was not aware the risk was 1 in 2 of transmission and cried when she learned this. She would abort an affected fetus. An experimental fetoscopy seemed the only choice. In spite of her chaotic childhood when her self-image was badly damaged by the defect (she used to keep her hands constantly in her pockets), her current self-image is surprisingly good, especially considering parental rejection. She only sees the defect as being a problem with how other people see her; she feels she can function very well with it, that it is not a real problem. But the problem of how she is perceived by others is one that she does not want her child to share. Here is another combination: positive self-valuation but negative social valuation. It is difficult to maintain a sense of self-worth in the face of continuous negative social feedback. We are not isolated oak trees able to withstand lack of any positive reinforcement. Our sense of worth is determined to some extent by the judgment of others. Self-worth alone may not be sufficient. Does choosing to abort affected fetuses result in a cumulative devaluation of individuals having the particular defect? Does it create a negative social consensus about such persons?

Case Study 64. Spina Bifida Pregnancy

An 18-year-old girl, now pregnant, had had surgical repair for spina bifida at birth. She was only mildly affected and functions well. According to her, she had been told by a physician that she would never become pregnant (this may be a case of patient-garbled information), and did not use birth control. The young father is not assuming any responsibility and the girl's mother is herself very immature. There are two problems. The girl has had no exposure to the variations in severity of spina bifida and does not understand the problems (physical and social) a pregnancy would mean for her. She would want to keep the baby, even if it had the defect (there is about a 1 in 30 risk). Her self-image, in terms of her genetic problem, has not been seriously devalued, but part of the reason for this is that her case is an unusually mild one. Experience of the other end of the range, although possibly beneficial in terms of exposing her to reality, might also be threatening if her sense of self-worth became linked to experience with the defect. If these two factors were not linked so intimately, then a decision about reproducing could be made with adequate, nondisturbing information and the decision would not involve self-repudiation.

Case Study 65. Achondroplasia
or Hypochondroplasia

A young couple came for premarital counseling, pressured by the boy's family. The girl is quite short, stocky, but with proportions not deviating extremely from normal. Her mother is even shorter, with the same body shape, and the pattern holds for her grandfather who was barely five-feet tall. Although the girl did not appear overly concerned, the boy's family was very concerned. A radiologist at first diagnosed her as achondroplastic and it then appeared that a classic dominant was the genetic mechanism, so that their risk would have been 1 in 2 for producing a child with achondroplasia. Prenatal testing is not available; however, a new diagnosis of hypochondroplasia, if correct, would considerably lower the risk and allow reassurance. One interesting aspect of this case was that although the boy's family had no qualms about insisting that the girl face her problem, they had managed to avoid indications of a possible chromosomal abnormality on their side. There was very high fetal wastage in the boy's pedigree (eight spontaneous abortions), but no evidence that his family wished to learn more about this. Again, since the expression of the girl's genetic problem appears mild, her only problem with self-image arises from a possible future family system that might have significant impact on her. We know that intimate systems, subgroups, and culture systems do make important value judgments of individuals. Should they make these evaluations in terms of genetic information? If they do not, can such larger systems produce meaningful ethical guidelines concerning reproductive choices?

As long as worth and reproductive choices are conceptually linked, the first premise (that individual worth should not be evaluated in terms of genetic diseases) will preclude ethical judgments about reproductive choice. But the implied assumption that choosing not to produce a genetically defective child means finding genetically defective individuals not worthwhile is, as I have indicated, a conceptual mess. Let me try to develop this in more detail.

We first assume that choosing not to have children, or therapeutically aborting a defective fetus, is the same as saying an existing individual is not worthwhile. It is easy to argue from that to the position that any form of dysgenics means judging some existent individuals to be unworthy, because it is blatantly circular reasoning when stated this way (choosing not to have defective children *means* that defective children are not worthwhile). And that is actually what assuming that individual worth is defined in terms of reproductive capacity and quality means in simple English.

However, recall my discussion about "possible individuals" in Chapter 3. What such arguments do is mix concepts of "possible individuals" with the realities of existing individuals, something no self-respecting philosopher ought to do, and something that makes no conceptual sense. I said in Chapter 3 that choices about "possible individuals" actually meant (in real life) choices made by actual individuals concerning their own *projects* (in the existentialist sense of that term), and that the ethical basis for these choices involved the individual's unwillingness to accept the pain and suffering involved in realistic projects, one of the consequences of our human ability to empathize. What the individual finds not worthwhile is the pain and suffering involved in the project. There is no other "individual" involved if we accurately analyze what in the world we could mean by "possible individuals." (I use "in the world" deliberately here. "Possible worlds" have the same status as "possible individuals": they are projects from an actual existence that need to be tested against reality as the process of experiencing continues.)

Thus, any reproductive choices involve the people who make them, and their attitudes toward pain and suffering. But people are social, are parts of systems, so that their choices indirectly involve all of us. That is the ethical base for judgments about reproductive choice, not an evaluation of individual worth.

Although this view may work well for carriers of genetic disease traits who are themselves normal, does it also work for the affected parent? Here is a real individual with the defect, and if we abort a fetus to avoid the pain and suffering we can project, does this apply to the actual person now suffering with the defect? What we wish precisely to understand here is what is meant by necessary and unnecessary pain. To some extent, this is still another choice we can make. Preconceptually, we tend to feel it is not necessary for us to experience the pain of projecting the development of a defective fetus. Prior to viability, depending on how we are going to view abortion, it is not necessary pain for us. But now the border grows shadowy. In the intensive care nursery, do we need to experience the present and future pain of the neonate? When does the living member of the human species become a person whose worth can be expressed in other than terms of biological existence? What I am going to suggest may sound overwhelming in its responsibility, but if we wish to keep from confusing levels of organization it follows. The terms "person" and "individual" are code words for a social level of organization, not a biotic one. As social terms, they are socially defined and their correspondence must be to social reality. It generally gives us great pain to kill a person, an individual (although there are major exceptions). If we balance the pain of homicide against the pain of empathizing with a very serious defect in an individual, the outcome is not in doubt in most cases, although it is in mercy-killing situations. But the definition of whom to consider a person or individual is to some extent a human choice.

Even here, however, the discussion has been in terms of social understanding of "individual" and experienced pain, not in terms of worth. It is not always true, for example, that committing homicide is equivalent to not finding worth. We speak about valuing human life and therefore interpret the taking of human life as not valuing it. This leads us into all sorts of puzzles however. Does the commander who orders his men on a suicidal mission not value human life? Does the jealous lover not value his victim? Does the elderly woman who helps her very ill husband die not value human life? Does the society that accepts *x* number of deaths from a necessary pollutant not value human life? Does a capital punishment state not value human life? Do the people on a life boat who decide to throw two people overboard rather than swamp the boat not value human life? These questions are not that easy to answer precisely because being killed does not always equal being of no worth.

Even if it did, however, to imply that preventing future pain implies devaluing individuals currently in pain would make no sense in terms of more routine Preventive Medicine procedures. It is only the weird blend of genetic problems and worth (as in psychiatric issues, the blend of a mind–body dualism and personal integrity) that allows this implication of devaluation more credence than it deserves. Imagine not removing a carcinoma because it would devalue those with inoperable carcinomas, not vaccinating because it would devalue those who already have the disease, or to be more fair and remove the element of survival, imagine not giving a terminal patient in great pain enough medication to remove the pain because the dosage might be a serious risk to him and we do not want to devalue terminal patients. The connection between genetic heritage and self-worth is difficult for me to see, and I can only explain its persistence by the analysis just offered. In discussions of genetic issues, we often hear a formula that reads: prevent the birth of defective "possible individuals," equate "possible" with actual individuals, and therefore desire to remove from existence defective actual individuals at all cost. Therefore, the argument continues, since removing from existence = not worthy, then we are judging such defective individuals to be without worth. That formula is filled with indefensible assumptions: (1) we prevent a projected future, not an individual, and do it on the basis of avoiding an increase of pain for us, not an evaluation of worth; (2) "possible individuals" are not actual individuals, not even by a stretch of the potentiality principle (which is demonstrably wrong in any event); (3) the cost of preventing pain must be weighed to ensure that less pain is generated by the act of prevention than is prevented by the effort; and (4) causing to go out of existence does not always mean holding one to be of no worth.

This flawed perspective in genetics issues is most commonly seen when discussing how genetic prevention will affect our perception of the children suffering from the various genetic defects, both in terms of the perceptions of the immediate family and of the wider social view of them. I would like to present a few case studies addressed specifically to this point, to see whether the empirical data correspond to my preceding analysis.

Affected Offspring

Case Study 66. Down's Syndrome

The wife is again pregnant and would like prenatal testing for Down's syndrome. They have a baby boy with Down's who is healthy, cheerful, and loving, and apparently very much loved in return. The husband is Catholic and the wife is a convert. They work very intensively with their son and are also very active in a Down's parents group. However, if the prenatal test indicated a Down's fetus, they would abort. Their son takes a great deal of time and effort, and they do not feel they could properly care for two Down's children. They talked calmly about their feeings about selective abortion while cuddling or playing with their son. They are pleased with the availability of the test, and had considered adoption as the only other alternative. There seemed no apparent connection in their minds between their son's worth and discussion of selective abortion for Down's.

Case Study 67. Encephalocele

The couple have a young son born with an encephalocele. Although the wife would like to have the test, the husband is opposed to it and his wishes prevail. He finds the small risk of the test unacceptable. Also, in his work he sees killing first hand and does not want to consider the possibility of killing the fetus. He sees the fetus as a person, in fact, in terms of his young son, so that killing the fetus would be like wishing his son's death. In this case, there is a connection made between the child with the defect and preventing the birth of another defective child.

Case Study 68. Congenital Chloride Diarrhea

The 10-month-old girl has been hospitalized with a rare recessive genetic disease, the thirty-first case in the world, and her long-range prognosis is guarded. Her mother and father are very young, and the mother is again 6 months pregnant. The grandmother handled most of the communication with the doctors. The mother cried frequently when informed of the diagnosis, and besides being upset, was angry at the situation and at not being given the diagnosis that it was a genetic disease before the second pregnancy. There is a 25% risk of another affected baby. Since they would like more children, but not more children as ill as their daughter is, it is a very stressful situation. The parents are very loving toward their affected daughter, spending much time with her, talking about her with pride, affectionately touching. But they also do not want another child with the disease and are very upset with the risk. Again, a connection between their perception of the worth of the affected child and their wish not to have more affected children does not seem present here.

Case Study 69. Thalassemia

There are two case studies previously discussed that are relevant here as well. One is the thalassemia major case. The couple have two children doing poorly (life expectancy is in the twenties), and the wife underwent fetoscopy to determine whether her accidental pregnancy would result in another child with thalassemia. They would have aborted if that were the case, because they could not bear to experience the suffering involved. They love their two ill children and did not see this decision as devaluing them in anyway. The second is the Lepore/thalassemia case. After their daughter was diagnosed, this couple decided to have no more children. An accidental pregnancy was aborted. The father shows considerable concern and love for his daughter (he too has the disease). The abortion was not seen as making any negative statement about his worth or hers. In both cases, the decision seemed rather to be based on the parents' perception of how much suffering they were able to bear, not on some evaluation of their own worth.

Case Study 70. Down's Syndrome

The couple have a 3-year-old with Down's syndrome as well as a 6-year-old daughter. The Down's child is in a special research treatment program. The parents feel they could not terminate a pregnancy if the test were positive for Down's because of their experiences with their child. However, they would consider terminating for other defects. In this case, some connection is again made between the reproductive decision and the affected child.

Case Study 71. Down's Syndrome

The 1-month-old Down's baby is doing very well. The parents seem to be adapting and are trying stimulation exercises to aid their child. They are not sure why they were referred for counseling. They know about the test and the recurrence risk figures, and if they accidentally became pregnant, would have the test done. They are disappointed, since they wanted another normal child (they have a 3-year-old). They also want both children to be capable of independence and separation from parents, and they are getting conflicting projections of the possibilities of this for the Down's child. The husband has decided not to have any more children.

This raises an interesting aspect of the "worth = reproductive choice" assumption. If this assumption were true, then would the choice not to reproduce because of the risk of a genetic defect devalue all those individuals who had the defect? In order to maintain the worth of human beings, would we ethically have to demand the actualizing of every "possible individual"? Although bizarre, this is not an unknown position in the discussion of reproductive choices. It seems to follow easily from the assumption of an identity between individual worth and the election to reproduce, and eliminates the possibility of any nonreproductive choice. Few of us would be willing to buy its consequences, but it is a reasonable inference from the "worth = reproductive choice" and "possible individuals" concepts. That alone should give us some humane pause about the true value of those concepts. To salvage the argument against selective abortion, it is possible to try and draw a distinction between killing a fetus and allowing the wastage of an egg and sperm, however. We can make the choice of the moment at which to prevent the process the crucial issue, and there *is* a difference, for many of our purposes and interests, in where and when we want to attend to experience. (If you want to build a house, you do not gather acorns—one of the reasons the potentiality principle is wrong.) There are more costs, more painful consequences, to stopping the procreative process at 5 months gestation than to stopping it with birth control measures. But does this help the worth argument? If a person's worth is to be judged by willingness to give birth to more individuals in his or her image, does it help to know at what point in the process we decide not to have more such individuals? Probably not. If we insist on linking worth with reproduction, we will devalue ourselves every time we decide not to reproduce, a very odd result from, of course, a very odd assumption to begin with.

The case studies presented here all indicate that the affected children are loved. There are, naturally, other cases where defective children may not be loved, but such lack of affection may have more to do with shame, guilt, anger, and withdrawal (nonrational affective responses) than with a calculus of worth based on supposedly rational reproductive decisions. Only one of the cases personified the developing fetus, and that personification was borrowed from the actual son, not developed from the possible personhood of the fetus (a phrase that seems meaningless to me). In all the cases, the concern was with pain and suffering, how to prevent it or how to cope with it. A choice was made based on the perception of an attitude toward pain and suffering, not on how the parents valued the affected child or children. In two cases, the parents decided that experience with their Down's babies had not produced more pain than they could cope with, and that they could therefore handle further experiences like this. In two other cases, the parents decided that watching their children's suffering with a serious blood disorder was the limit of their ability to endure their own pain, and they could not adapt to additional painful experiences. In neither case was the worthiness of the affected children evaluated or changed by the decision. The pain was. Which brings us back to our interventionist discussion. The justification of intervention, based on attitudes toward pain and suffering, is the ground for the ethical justification of prenatal interventions, as for all medical specialty interventions. And on this, as I have said, there is a serious and fundamental split between the sciences and the humanities that underlies and often makes unintelligible much of the medical ethics debate. Where the debate tends to focus unfairness is at the balance sheet: the costs of intervention are usually added up in detail; the costs of non-intervention rarely are acknowledged. Fundamentally, however, the bottom line is what our attitude is toward the pain both types of cost represent.

One last problem about the responsibility of carriers, especially as it affects the care of children with defects. Instead of discussing patients' "rights," I have preferred to call them interests and needs. There is another side to the coin: duties, obligations, responsibilities. Where do these fit in? If one allows "rights," of course, obligations follow pretty straightforwardly as necessary to secure or insure rights. But I have said there really is no adequate conceptual base for a "right"; it is a concept in mid-air. Therefore, responsibilites (which I like better than "duties" or "obligations" for its social rather than legal sense) must arise from interests and needs, and they do easily enough. They come from our interests and needs vis-à-vis other people and from our interests and needs in self-development and self-satisfaction. We need to assume responsibilities to create intimate social networks and to create broader social networks. We need to assume responsibilities in order to function in increasing numbers of ways, to expand ourselves. Assuming responsibility in general means accepting the consequences of our actions, anticipating and choosing those consequences, and then incorporating them into our life pattern. There are some social roles that ask little of this, but most require it. A list of exceptions gives a good feel for the self-development implied in accepting responsibility: roles of very young children, prisoners, very sick individuals (physical and mental illnesses), and very old people (if they wish). The very powerful, although they have the means to avoid most social consequences, by definition cannot avoid the major consequences: they are in charge.

If we do not wish to assume responsibility, then, we have few social roles available to us. In some way or other, most of us have to choose and live with consequences. Carriers for genetic problems, unless they fit one of the few nonresponsible social roles, must therefore make a choice of consequences. Some have argued that living with the unpleasant consequences of that choice should not be required,[44] and this is a difficult problem. Not to require it is to create an additional nonresponsible role, closely akin to the sick, immature, or specially privileged roles already available. To require it is to allow the pain of the choice its full expression. In the former, we subtract from the individual's development and pride, and increase pain for others who must assume some of his consequences. In the latter, we subtract from our own compassion (empathy) and human warmth. I do not think there is a solution without heavy cost.

To some extent, the tragic choice will be made in terms of the severity of the consequences and our perception of how seriously self-destructive the choice may be (in which case, society may place the individual in an incompetent role and thus solve the problem). Social decisions may also be forced by survival circumstances: scarce resources, civil disruptions, natural disasters. When the social system itself is fundamentally threatened, more demands are made on individuals to handle the consequences of their choices without the buffer of social support. There is some question, in any event, of just how effective in this task social serv-

[44]See, for example, Daniel Callahan, "Health and Society: Some Ethical Imperatives," *Doing Better and Feeling Worse: Health in the U.S.*, John H. Knowles, ed., New York, 1977, pp. 23–34; "The Meaning and Significance of Genetic Disease: Philosophical Perspectives," *Ethical Issues in Human Genetics*, Bruce Hilton et al., eds., Plenum, New York, 1973, pp. 83–90.

ices really are. One of the things we really have to determine is whether an impersonal service system *can* assume the consequences of individual decisions, and this has to be answered from two aspects: the buffering of painful consequences for the person who made the choice, and the alleviation of painful consequences to those others affected by the choice. Another issue to investigate would be the effect of converting direct individual responsibility to indirect institutionalized responsibility. I do not propose to attempt either task in this book. Where these issues impinge on reproductive choices, they do create significant problems and I think I have indicated that the answers are not easy.

If carrier parents choose to have a Lesch-Nyhan child, consequences are incurred not only by them (constant supervision and care activities, medical expenses, change of personal relationships), but by the child (pain from self-mutilation, confusion, difficulty in communicating, poor physical prognosis, uncontrolled activity), other family members (concentration of time and attention on the affected child), and the larger social system (shared medical costs, distribution of medical resources, preventable increase of perceived pain). Such choices put society in a no-win situation, a true dilemma. Once the choice is made, there is no solution without overly high costs, either to the parents, the affected child, or the extended system. To wish these costs away by attending only to the value of free choice and overlooking the value of responsibility (accepting the consequences) with all the resulting issues left unmet, is a dangerous denial of reality. If we waive such responsibility for carriers, the problems I have discussed still must be handled somehow. We also need to justify such a waiver ethically, and if we do it on the basis of an autonomy value, we must make sure we have considered other values in the equation and have a clear idea of what autonomy actually means for us. Is one autonomous, has one really made an independent choice, if the consequences for that choice do not have to be personally assumed?

In addition to the value of autonomy in an ethical system, there are values of compassion, social relating and caring, well-functioning, and so on. These choices frequently occur in the context of conflict. The argument, however, that carriers should be autonomous and nonresponsible is a peculiar one in ethics, and adds to the confrontation. When we wish to broaden the concept of nonresponsibility to mean that an individual's choice, no matter what, must be ethically approved, the full weight of the argument must rest on an overriding valuing of free choice and the ambiguity in the meaning of free choice. The flaws in this should be apparent:

(1) There is no good reason to make free choice a value that overrides all others. We do not do it for many people who choose to kill or steal or slander. Why should it override compassion or reality testing or denial in all situations?

(2) What sense can we make of free choice when we assign the chooser to a special class of nonresponsibles, thus practicing a paternalism that removes all meaning from that freedom of choice? We contradict the overriding value of free choice in a subtle way, since, if whatever we choose is by definition right, it makes no difference what we choose, our choosing may as well be random, and therefore nothing is contributed to our individuality or autonomy. Our free choice is unimportant as free choice.

On the contrary, then, we can make ethically negative statements about the reproductive choices of carriers: some will be good choices, some bad. The evaluation needs to be based on consequences to the chooser, to the affected children, to the family, and to the social system. How much preventable and unnecessary pain has been added for those involved because of the choice? Is it more than would have been entailed by making other choices? We therefore need not put carriers in a special class of nonresponsibles when they make reproductive choices, at least in terms of our ethical evaluation of those choices. Whether we should put them in that class in terms of buffering them from the consequences of their choices by socially or economically supporting those choices is another question that I have indicated has no simple or pleasant answer.

I want also to deal with the issue of guilt or shame, of personal responsibility, not for reproductive choices, but for simply being a carrier of a genetic disease, or of feeling somehow responsible for a birth defect without ever having made a reproductive choice concerning it. In terms of accepting the consequences of actual choices, this is a quite different sense of responsibility from what we have been discussing, and needs to be handled in some detail.

Parents' Sense of Responsibility
for the Defect

Case Study 72. Developmental Delay

The parents came for reassurance. Shortly after birth, their baby was found to be hypotonic, with an inadequate sucking reflex and increasing apnea, and was also diagnosed as having cerebral disorganization. Now, at one year, the child is active and has good language development. A chromosome analysis was normal. During counseling, the mother expressed her concern that the emotional stress she experienced during her pregnancy might have caused the perinatal problems. She cried when reassured that this was not the case, since this reassurance removed the blame she had been assuming. Often in a counseling session, one or both of the parents are carrying concealed burdens of guilt. They assume an unwarranted responsibility for events that were actually not under their control, owing to their lack of information or magical thinking. Some of the shame linked to genetic illnesses is generated by this sense of being at fault, and this fault is handled in the same way as conscious choices or actually controlled events.

Case Study 73. Meckel's Syndrome

Because of an ultrasound scan that indicated an encephalocele, and very high alphafetoprotein levels, the parents decided on abortion. This was their second child with Meckel's syndrome, and they had been very carefully counseled about the genetic mechanism of the disease. They knew it was an autosomal recessive and understood the information. Nevertheless, when discussing the autopsy report with the physician, the mother asked whether the baby's problem could have been caused by the mother's uterine structure, or by getting pregnant too soon after the previous pregnancy. The genetic problem was still viewed as a possible personal fault or a general body defect, not as a specific mistake in DNA over which the mother had no control. Rather than investing responsibility in the reproductive decision, the mother is placing responsibility in her physical self (with the usual effect that has on feelings of self-worth) and on mistaken decisions about the timing of the pregnancy (being overeager, and hence psychologically devaluing herself). We now have a problem with her exaggerated assumption of personal responsibility.

Case Study 74. Hunter's Syndrome

A 16-year-old boy is institutionalized with Hunter's syndrome. He is constantly drowsy, non-ambulatory, has respiratory problems, and is deteriorating. The specific issue is to help his younger sister by determining her carrier status. The sister is very concerned, and had decided not to have children before she was informed of the availability of carrier and prenatal testing. She can be aided by genetic counseling. Hunter's syndrome is X-linked, and the extended family is very bitter about this disease. The grandmother has guilt feelings because both her daughters are carriers and she feels responsible as a person.

Genetic diseases are not the only cases of such exaggerated feelings of responsibility, of course, but this is a frequent characteristic of theirs. It involves both a misperception of what is and is not under our control, and a confusion of person-level with physiological-level. It also reflects a concentration of attention on the reproductive parts of the body, valuing them more highly than other body subsystems. There is certainly some reason for this when we are interested in issues of evolution, and I don't wish to imply that this is not an important interest. Species adaptation and survival, as well as individual adaptation and survival, is a primary value for the affirmative stance, as I have stressed. However, even on the reductive physiological level, and certainly on the social person level, it occurs in the context of other choices and values. The body is not simply a way of continuing the egg; that is a very simplistic notion. The body in the changing environment is an experiment in process that does not continue the status quo of some archaic egg cell. Rather, variations result from the transaction, which involves the adaptive history of the whole body to changing external conditions. There is no inner drive to continue an insulated and nonreactive germ cell. There *is* a drive to continue the process of adaptation, which depends on the entire functioning body. Predicting which genetic variations will assist that adaptation is not a precise science, and with the presence of change, may even be theoretically fruitless. It would call for information on the reductive level that we do not have, and may never sufficiently accumulate, for equations to manipulate the range of interactions that might simply be impossibly unwieldy, and for a detailed determinism that could accurately predict the actual moves and countermoves of adaptation.[45] We can make some rule-of-thumb evaluations, however, and I may not be quite so pessimistic as Edmond Murphy about our ability to evaluate some genetic components in terms of evolution.[46]

In any event, on the person level, there is a limited set of behaviors that could be viewed as conferring culpability for genetic

[45]See my "Stasis: the Unnatural Value," *Op. Cit;* Ludwig von Bertalanffy, *General Systems Theory*

[46]Edmond Murphy, "Genetic Engineering: Eugenic Consequences," *Science and Morality,* D. Teichler-Zallen and C. Clements, eds., Lexington Press, Lexington, Mass., 1982.

problems on the individual (e.g., taking mutagenic drugs unnecessarily). In most cases, the individual could not have altered the outcome (assuming no knowledge of personal carrier status) by any alteration of personal behavior. The gene for Meckel's syndrome is not under anybody's control. Reproductive choices concerning that gene are. But simply having the gene is not that person's responsibility. The eagerness with which individuals accept responsibility, even for very unpleasant events, says something about either our need to feel or our habit of feeling that we are in control of events, a very ancient and continuously expressed aspect of human character. Even in the face of a massive burden of guilt, we prefer the feeling of control of our lives to the fear that important things can happen to us without reason and without our wish. The ideal of creating a fair universe, or of a "just realm" beyond, is one expression of this. That genetic roulette could cause a child of ours to suffer and die is a threat to our affirmation of reality, and hence to all our values and meanings. In the face of this threat, searching for some fault of ours, something we did, is a painful but workable defense. However, the cost of such defenses is usually too high, being unnecessarily self-destructive when other coping mechanisms are possible. The genetic counselor can usually justify intervention to remove such unwarranted guilt.

Summary

I think it's clear that personal worth must be conceptually separated from reproductive capacity, although it is also clear that this is a culturally ingrained confusion. The linkage has generated a number of untenable positions in medical ethics discussions of genetic issues, and made most analysis a muddle. If we can free concepts of individual worth from reproduction, we can then engage in the required analysis of the components of personal worth that I outlined in the opening pages of this chapter without raising a genetic threat to self-image. I plan to look more closely at these components of worth, and how they relate to the abortion/infanticide issue, in Chapter 6. For now, I will be satisfied to have disconnected worth from reproductive capacity, and from killing in certain situations, which is a necessary preliminary.

Physiological abnormality, then, does not equal "fault," and the choice to abort does not equal finding a person or fetus with such an abnormality "nonworthy." The choice to prevent pain, to choose against a projection, is a more useful explanation for such choices than devaluation of a possible individual. From an ethical perspective, such choices can be seen as responsibilities (in a social context) that are generated by our needs and interests. This notion of responsibility also involves living with the consequences of our choices. The ethical basis for any such evaluation of choices is not the traditional notion of free will, but rather, an affirmative attitude toward existence that imparts a responsibility for well-functioning.

Chapter 5

Experimental Research and Procedures

There are two fundamental types of research: (1) Therapeutic (clinical) experimentation, which is research aimed at benefiting the person on whom an experimental procedure is tried by mitigating an actual problem suffered by that person. (2) Nontherapeutic (nonclinical) experimentation, which is not intended to benefit the subject directly in terms of the specific problem the experiment addresses. It is often assumed that nontherapeutic experimentation conveys no benefit to the subject, or the benefit is lost sight of in discussion, because the experimental problem is not one the subject shares. Thus testing the saliva of healthy parents and babies exhibiting no genetic defect or risk for myotonic muscular dystrophy, in order to develop procedures for testing for muscular dystrophy in families at risk, conveys no possible therapeutic benefit to those without muscular dystrophy. But this is only one possible way of categorizing experimentation and it narrowly limits justification of clinical experimentation to studies carried out to determine the side-effects of the experimental therapy, the prognosis for the problem if untreated, the risk-taking attitudes of the subject, or the possible benefits if successful. The problem with the therapeutic/nontherapeutic classification lies in thinking that nontherapeutic experimentation must be justified in the same overly narrow terms as therapeutic research. It is easy to limit clinical research to a narrowed cost/benefit analysis based on the notion of therapeutic intervention or of a partnership with the subjects in the battle against the disease that afflicts them, since such research can generally be easily justified in this way. But reasoning in these therapeutic terms can, however, seriously confuse the is-

[47]Leroy Walters, "General Issues in Experimentation," *Contemporary Issues in Bioethics*, T. Beauchamp and L. Walters, ed., Wadsworth, Belmont, Ca., 1978, pp. 399–403, for example.

sue when thinking about nontherapeutic experimentation since these reasons remain only part of the justification, that part specific to the therapeutic characteristic of the research, and should not be confused with the broader ethical justification of research in general. It is the broader justification that nontherapeutic research depends on more heavily, and that remains the keystone for *both* subsets of research. We have often misplaced the emphasis, stressing the characteristics of the subset of clinical research as the ethical ground for all research, and hence talking about risk–benefit ratios for the patient in terms of the specific illness, and about utilitarian calculations of disease group benefits vs individual health risks, and about investigative covenants between the ill patient and the healing researcher. We need to go back to the basic ethical justifications of general research: natural and nonrational expressions of empathy, the social nature of human beings and their interest in and responsibility for a social system of which they are an intimate part, and the reasons for attempting intervention in situations (in this case, experimental intervention). These justifications are the same for therapeutic or nontherapeutic experimentation.

Although some nontherapeutic research is conducted in our program, the case studies will concentrate on therapeutic experimental procedures. However, the discussion really does not differ that much for either in terms of their basic justification, as I have indicated and will develop during the presentation of these case studies.

Thalassemia Research

Case Study 75. Fetoscopy

In the last chapter, I discussed briefly the case of a couple who had two children ill with thalassemia and who had to make a decision on a pregnancy resulting from failure of a birth control technique. They did not wish to have another child with thalassemia and would abort if prenatal testing could not be done. Currently on the

[48]For examples, see Hans Jonas, "Philosophical Reflections on Experimenting with Human Subjects," *Contemporary Issues in Bioethics*, T. Beauchamp and L. Walters, eds., Wadsworth, Belmont, Ca., 1978, pp. 411–420; Paul Ramsey, *The Patient as Person*, Yale Univ. Press, New Haven, 1970; Richard McCormick, "Proxy Consent in the Experimental Situation," *Contemporary Issues in Bioethics*, pp. 456–465.

East Coast, Yale is doing experimental fetal blood sampling using fetoscopy. The risk to the mother is no greater than routine amniocentesis, but fetal risk is higher. The sample is small, but based on over 100 fetoscopies done, the risk figures are: 5% spontaneous abortion, 8–10% premature delivery. They decided to try the procedure, but the first test results were equivocal because of maternal blood contamination of the obtained sample. The wife was given transfusions before a second test to attempt to preclude this. The couple returned for a second fetoscopy, expressing a very touching view of the circumstances: "The dance once begun must be finished." The sample analysis showed the fetus was heterozygous. A few months later, the couple had a healthy baby boy, full term and uncomplicated. The experimental nature of this prenatal test had been carefully explained to the couple and the physicians were nondirective concerning it. Abortion or the risk of another affected child were the only alternatives. The parents made the choice to intervene experimentally in the pregnancy on the basis of their experience with the suffering of their two children and their unwillingness to allow additional, similar pain as a future possibility. The procedure was experimental research in terms of the parents' needs and the wife's pregnancy. The fetus was a subject in a very secondary sense at this point, and it might more accurately reflect the situation to consider it in terms of a developing pregnancy process whose outcome had not yet been decided. Of course, any experimental procedure that terminated 75–100% of pregnancies would hardly be a medically acceptable prenatal test, so that status of the developing fetus is germane to the evaluation of the procedure. But it is germane primarily in terms of the interests and projects of the couple involved, not in terms of the fetus as a person, and hence not as a subject of research to be protected by the ethics of research. This concept of person-level and harm to embryo or fetus will be developed in detail in the next chapter.

It should be stressed that experimental fetal blood sampling will be justified or not on the basis of the general ethics of intervention: (1) Is this a situation in which a medical intervention can prevent pain and suffering, or is this necessary pain and suffering? and (2) Is this a situation in which experimental intervention, and hence higher risk-taking, has some reasonable chance of success and occurs in a context of sufficiently high stress or risk for the subject to justify the experimental procedures? In addition, (3) Will taking part in this research allow the subjects to express social aspects of their persons in constructive ways, give them a sense of community and meaning that will contribute to their development? If, for example, the test had caused a spontaneous abortion, could it still have had a purpose for this couple in terms of giving them a way to contribute to understanding of the procedure and giving their choice a larger social context? Finally, (4) Will taking part in this research actualize the couple's empathy with those not only in similar circumstances, but in varying circumstances in which medical intervention may be a possibility?

Case Study 76. Chelating Pump

I have also previously discussed a Lepore/thalassemia case in which the daughter has been in our hospital for frequent transfusions and might be a candidate for an experimental treatment. These patients usually die of iron overload, since iron buildup damages the heart and liver. One of the costs of the transfusion treatment is that transfused patients develop an overload sooner. There is a new treatment with a chelating agent (Desferal); however, it is not effective orally and an 8-hour per day transfusion is presently the only workable method. Home trials have been effective and the hospital is also considering offering the treatment. Six affected families have met to discuss the new treatment but were somewhat depressed by what it involved. Although the pump is expensive ($1000) and drug cost is also considerable ($4000), the National Institutes of Health will cover the cost for those who join the research program. The families were given three articles explaining the new treatment. Although the daily treatment would remind the family each day of the seriousness of the illness and emphasize medical dependency (psychiatric experiences with renal dialysis machines are not encouraging[49]), long-term prognosis with current standard treatment is grim. The cost of such hypertransfusion treatment is also quite high ($10,000–20,000). However, the chelating treatment is indefinite, with no end in sight. The six families reached different decisions: three were not interested in the new treatment, two wanted to take part, and one could not reach a decision.

[49]Harry S. Abram, "The Psychiatrist, the Treatment of Chronic Renal Failure, and the Prolongation of Life, I, II, and III," *American Journal of Psychiatry* **124**, 10, Apr. 1968; **126**, 2, Aug. 1969; **128**, 12, June 1972; pp. 1351–1358, 157–167, 1534–1539; Franz Reichman and Norman Levy, "Problems in Adaptation to Maintenance Dialysis," *Archives of Internal Medicine* **130**, 859 (1972).

In this case, the decisions concerning the research were again based on general considerations concerning intervention: the massiveness of the medical intervention (8 hours a day, every day, indefinitely) seems to have been a major consideration and the projected effects of this intervention outbalanced even the projected early death of the children. Had the parents the right to make these choices for their children, either affirmatively or negatively? What is the ethic of the surrogate role? If Ramsey were right that we could not choose to volunteer children for nontherapeutic research, and that the basis for this injunction was that children cannot enter into the "consent as a canon of loyalty/adventurers in a common cause," he is wrong that therapeutic intervention could be justified either. Conceptually, whether the intervention is therapeutic or nontherapeutic is irrelevant in terms of consent and covenant. If a surrogate is impossible for one, it is impossible for the other; if possible for one, it is possible for the other. Therapy makes no difference if the ethic is based on capacity for consent. This is a consequence I certainly can appreciate Ramsey's not being willing to accept, but I think he needs to conceptually in order to be consistent. The ethical justification for research is not based on consent and covenant, however, although consent is an important component for those capable of consenting and surogate consent for those not. A component and a base are quite different.

Case Study 77. Research on Genetic Counseling Methods

One of the objects of this study was to obtain a volunteer population not self-selected because the research was relevant to their genetic problem: the reason for consenting to become a research subject was not for therapeutic benefit. Through a health care organization and with emphasis on preventive medicine, a thalassemia screening program was conducted. Patients understood that multiple screening would be done routinely. Identified thalassemia carriers were sent a letter requesting they make an appointment to receive information concerning lab results. When they arrived, and prior to counseling, they were asked if they would consent to participate in a study to determine the clearest way for physicians to present medical information to patients. It was made clear to them that whether they offered to participate in this research or not, they would receive the same information about their lab tests. There was a 91% consent rate. The purpose of the research was to help determine the best genetic counseling method (the most economical way to obtain an acceptable result), and the subjects had been told enough about the study to think it would be of general benefit to patients, but did not know how specifically it would benefit others with a problem like theirs. Besides attempting to evaluate videotape counseling in terms of traditional physician counseling and controls, the study investigated mood changes that might result.[50]

This research was not physiologically risky for the subjects, although it did involve a fair amount of their time and would effect some psychological changes. The high participation rate is interesting because as far as the subjects knew, it was nontherapeutic research. In this case, potential therapeutic benefits were certainly not the motivation or justification for participating. Rather, a more complicated combination of social and intrapsychic factors were at work, as nicely outlined by Frank Ayd, Jr.[51] Nor were the subjects completely informed of all the goals of the research, or of the research design. To have done so would obviously have invalidated the results. Thus, consent and covenant were also not the primary justifiers. So far, I have been describing this work sociologically or anthropologically. The question is, *should* these social and intrapsychic factors have been the justification, or were the investigators obliged to appeal to therapeutic risk/benefits or consent/covenant in order to justify use of the subjects.[52] Both these approaches make the grounding of ethics dependent on rational elements and overlook important nonrational elements in their ethical scheme (Ramsey manages this by concentrating on the informing function and the rational decision-making process in which even covenant becomes an intellectual rather than a social adventure). If we ask why, fundamentally, we do research, we get very different answers. Research aims to give us more potential for intervention, or to allow us to choose the right intervention to achieve our purposes. It is first justified in terms of our attitude as interventionists rather than as accepters, which is a nonrational

[50]Peter T. Rowley, Lawrence Fisher, and Mack Lipkin, Jr., "Screening and Genetic Counseling for β-Thalassemia Trait in a Population Unselected for Interest: Effects on Knowledge and Mood," *American Journal of Human Genetics* **31**, 718 (1979); Lawrence Fisher, Peter T. Rowley, and Mack Lipkin, Jr., "Predicting Immediate Outcome of Genetic Counseling Following Genetic Screening," *Social Biology* **26**, 289 (1979).

[51]Frank J. Ayd, Jr., "Drug Studies in Prisoner Volunteers," *Southern Medical Journal* **65**, 440 (1972).

[52]Henry K. Beecher, "Ethics and Clinical Research," *New England Journal of Medicine*, June 16, 1966, pp. 1354–1360.

base. It is also legitimized because we can empathize with others as well as have interests and feelings about ourselves. Compassion, caring, or empathy is also a nonrational factor. If we ask why we choose to participate in research as subjects, the answers follow similar lines. We wish to help in the process of changing, rather than accepting, circumstances. We feel a sense of responsibility to engage in social dynamics of this sort, because being a social species, we have a need to feel part of social processes and this is one way to express our social selves. This is not a moral imperative "ought" or a required legal obligation "ought," but behavior based on our sense of belonging. If we are sufficiently alienated because of mistakes in the socialization process, there is no guarantee that moral duties or legal obligations can create the social glue to hold groups together. The center will not hold because the sense of personal empathy and belonging are missing, non-expressible. Further, the social system that does not provide sufficient channels for expression and enhancement of this behavior accelerates alienation. Offering to participate in research not only enhances the social development of individuals, but also contributes a sense of meaning and an expansion of empathy that involves individual growth as well. The ethical criticism of research sensibly rests on its design, purpose, and meaningfulness. A poor study, a useless one, an overly duplicative one, one that falls below minimum scientific standards, deprives subjects of all purpose and meaning, cheats them of a part of themselves.

Finally, critics of research practice a paternalism in attempting to decide for individuals what research they may volunteer for or in trying to assume a guardian role between parent and child. In *exceptional* circumstances, this is ethically justified in spite of the cost to individual development, responsibility, and uniqueness. In the first case, if we see the individual as clearly incompetent and/or in a highly controlled situation under powerful conditioners, the goals of development, responsibility, and expression of empathy make little sense. But we need to be careful that our judgment of the situation corresponds to the reality of the situation. Prisoners *qua* prisoners, for example, are not automatically incompetent to make whatever choices they see open to themselves, and/or are not automatically controlled and conditioned as in the total institution model. Potential subjects may not need to be totally educated to the research protocol in order to understand enough to decide whether they would like to take part in research as a way of channeling empathy and social feelings, and/or the persuasive power of the medical degree or the institutional symbols does not *prima facie* outweigh all other heterogeneous conditioning factors human beings are exposed to.

In the second case, attempting to replace the parents in the parenting role, to assume we can institutionalize the social bonds of the parent/child relationship through guardianship, ward of the court, or various social service replacements, is a risk justified only by an extreme breakdown of those bonds. I maintain this because we have psychological data on the importance of those bonds, and we do *not* have the outcome studies to indicate an adequate success rate for their replacements. In most cases, we trust that parents (or other members of the family system) will make choices concerning their children's participation in research based on how they project positive experiences for their child through such participation, with values such as contribution and sense of meaning being definitions of "positive."

Other Applications of Fetoscopy

Case Study 78. Ectrodactyly

This case was initially presented in the last chapter. The woman has ectrodactyly, was pregnant, and had a 50% risk that her baby would also have the defect. The only prenatal testing possibility seemed to be fetoscopy. She traveled to an east coast medical center specializing in this procedure, which is more invasive than amniocentesis and hence more risky. Ultrasound is also used by the local center in connection with fetoscopy, and is a considerably less risky procedure. Through very careful ultrasound scans, the physicians were able to visualize the fetus' hands and see distinct phalanges. Since the ultrasound results so clearly indicated a normal hand development, fetoscopy was not done. The woman was assured her baby would have normal hands and feet. She apparently still worried, however, having nightmares that her baby would have the deformity. Pregnancy went to term and she delivered a normal baby she describes as beautiful.

Fetoscopy can also be used for fetal blood sampling for hemophilia and sickle cell anemia. In both problems, alternative prenatal tests were sought, because of the invasive nature and risk of fetoscopy. For sickle cell anemia, research in recombinant DNA has now supplied the desired alternative. It is possible, using restriction enzyme DNA techniques on cultured amniotic fluid cells obtained by amniocentesis, to test for sickle cell anemia with 90%

accuracy. This could make testing more generally available than fetoscopy could be for a genetic problem with high incidence in a particular population.

Case Study 79. Hemophilia

The wife, 18 weeks pregnant, had learned a year ago that her cousin was a carrier for hemophilia. This was a wanted pregnancy with a good possibility of twins because of the use of a fertility drug. There were alternative options. A specialist could test her for carrier status with 94% claimed reliability, but they would have to travel to another state. A tap could determine fairly quickly the sex of the fetus, and if female, there would be no problem. If male, this procedure could not help. The couple could journey out of state for a fetoscopy and test of the fetal blood sample obtained, but only four had been done and results might owe to chance. They were not considering aborting. They decided to try both carrier testing and amniocentesis. Ultrasound indicated twins and the tap showed both were male. Carrier testing was done, and fortunately determined the wife was not a carrier. If she had been, the next level of treatment would have been fetoscopy. In view of their decision not to abort, would such an experimental procedure have been justified? Even if the patient was eager to take part in such therapeutic research, should the researchers ethically allow it? Which value in the situation should rank highest: the subject's wish to contribute to research to develop successful medical interventions involving a shared problem, or the risk/benefit ratio for the patient? Who should make the decision?

Case Study 80. Duchenne's Muscular Dystrophy

Fetoscopy and the CPK test have been suggested experimentally as prenatal tests for Duchenne's muscular dystrophy. The couple being counseled had managed to accumulate a great deal of misinformation about their problem, either through misunderstanding or selectively hearing their physicians' explanations, or through being actually misinformed. They came with high expectations, thinking amniocentesis could detect DMD, which is not the case. The wife had also originally been told that her chance of being a carrier was astronomically low because of the number of healthy males in her family history. When she was tested, however, it indicated she was a carrier, but the test reliability is now estimated at

70%. It was necessary to review with this couple the uncertainty of our knowledge. For example, since CPK is not the direct gene product, the test validity is a problem. CPK levels can be elevated for a number of reasons and false positives on CPK have been reported in *Lancet*. The capping theory may or may not be adequate. Amniotic fluid cannot be used to determine whether the fetus is affected. Experimental fetoscopy is a risky procedure that, in the light of questions about CPK test validity, may be difficult to justify. The couple decided to use amniocentesis for sex determination. If the fetus is female, there will be no problem, but a male is at 50% risk of having Duchenne's. The physician did not encourage fetoscopy. Should he have?

I discussed the limits of patient decision in terms of the expert role in Chapter 2, and much of what was said there applies to justification of experimental procedures in terms of the patient's interests. The patient's perceived interests are part of the medical transaction, and contact with the physician implies that the patient has chosen intervention over acceptance of a continuing situation. However, in the expert role, the physician adds additional considerations to the decision-making process. The patient's interests may be evaluated by the physician as unrealistic or even self-destructive in light of the physician's expert analysis of the situation. That analysis can vary considerably in its correspondence to intersubjectively agreed-upon reality. Certain physiological data, certain scientific theories, are probably well in agreement, if the physician is competent. In these areas, the physician might even make unilateral choices. Fetoscopy will not be done in the Duchenne case because the physician presents information to the patient that indirectly (or sometimes directly) recommends against it. The research group in the ectrodactyly case decided that the data from the low risk ultrasound procedure was adequate and chose not to do the much higher risk fetoscopy procedure. Even if the patient had demanded the procedure, in the expert role the physician may unilaterally decide not to perform it.

Similarly, the researcher can make choices concerning a subject's wish to participate in a project based on self-destructive motivations or excessively high risk factors, and we would expect a doctor to make such choices as an expert. Although the subject might wish to engage in high risk-taking behavior, the subject's autonomy does not imply that the researcher has an obligation to collaborate in that behavior. There is justification for intervening in self-destructive behavior at a number of levels.

On the other hand, should the researcher always emphasize the risk or preclude certain groups (such as the sick, prisoners, children, mentally incompetent) from taking risks at all? It is important to maintain context in these analyses. "Risk" is a judgment, made in terms of other routinely assumed risks, and can be distorted in either direction if not somehow ranked. Certainly unnecessarily increasing risk to the sick beyond a certain level conflicts with their interests and the physician's interest in their return to well-functioning. Any research that significantly jeopardizes that goal violates the nature of the physician/patient interaction. But is there any reason to exclude the sick from research in which no significant jeopardy exists if they wish to participate in such an activity? There can be only if we view the sick role in a particular way: the sick are conditioned by the hospital environment, or physician's office environment (a total institution) to respond affirmatively; the sick, because of pain, lack of energy, or feelings of dependence, have lost their ability to make truly individual choices or determine adequately their best interests; the sick need to be cared for and protected, using a children's model, because they are sick. I think there is some truth to all these characterizations, but I think they are not as unique to peculiar roles like sick roles as we would like to think. We are all conditioned, that is, we all learn from our environment and internalize some of those experiences. Before the sick ever reach the medical environment, other pluralistic environments have made them the individuals they are and to assume they quickly and readily lose that individuality when surrounded by health care providers gives more power and monolithic structure to conditioning than it deserves. The truth lies in the result of constant and reasonably long-term exposure to a monolithic medical environment, a true total institution. But this is not a characteristically occurring situation in the sick role. Most sick roles are transient, involve other varying experiences, or are broken up in time.

Pain, energy deficit, and feelings of dependence are common to us all, but become more intense in the sick role. Dependence can, in fact, represent a hindrance to natural recovery. Participation in scientifically proper and noncompromising research can, actually, reinforce the individual's personality and prevent restricting of autonomy. It is only if we assume that the sick role already equals complete dependence and lack of ability to make decisions that we can consider it a special category requiring nonparticipation in research. Increasing dependence is not improved by restricting choices. But, if dependence *has* reached the level where decision-making as we understand it in every day experiencing (not, however, as understood as some non-influenced rational ideal) is impossible, than either such sick individuals cannot be asked to participate in research, or surrogate permission is required. We would need to determine empirically that dependence had reached such a level, and not theoretically assume it must have reached that level by definition of the sick role.

Finally, "caring for" tends to be expressed in terms of a parent/child model that, in light of our past experiences, is a normal response. The sick role has been shaped by this model, but we need to remember that the sick have not abandoned all other social roles, except in extreme cases. Expression of some of the other roles may be curtailed, but individuals still define themselves in those terms. Most of those roles do not follow a parent/child pattern and the paternalism of the sick role need not extend conceptually to the other roles. Again, empirically, if "caring for" has actually removed all possibility of meaningful individual choice, then those in this sick role cannot be used in research or a surrogate must make the choice for them.

The same analysis can be done for prisoners. How monolithic is the conditioning of the prison institution, how much dependence and erosion of individual possibilities has occurred, does the paternalism (in this case, punishing rather than caring for) preclude expression of any other model? In other words, how much does the prisoner class vary from average individuals? In view of the *sub rosa* inmate governance system that flourishes in prisons despite the officially imposed system, I think we can question any assumptions about the power and efficacy of official prison conditioning. The latitude for individualization seems *prima facie* broad, so that prohibiting prisoner volunteers for research may in itself suppress opportunity for expression of that individualism. We may be conceptually turning things upside down.

Children and the mentally incompetent differ from the sick and prisoner groups by the very fact that we consider them incapable of making meaningful and responsible choices by definition. It is here that the issue of surrogates becomes important. I want to look at this more closely after considering a few more cases.

HLA Linkage, Secretor Linkage Research

Case Study 81. Cerebellar Ataxia

A 65-year old man was brought in as an inpatient because caring for him at home was too difficult. He suffers from a cerebellar ataxia that involves progressive loss of coordination, and in some types, loss of thought processes as well. His symptoms began in his forties, and his wife is sufficiently distressed by the effects to ask the physician to counsel her children not to have children themselves. This is a dominant and has affected six members of the family in three generations. One form of ataxia has demonstrated HLA linkage. If the linkage relationships could be worked out, knowing the HLA typing of an individual with the disease and that individual's parents could enable us to predict whether a sibling or children of the individual would be at risk for developing this late-onset genetic problem, and might even give possibilities for prenatal testing. It was suggested that HLA typing and research might be done on this patient and his family. Such research could not therapeutically benefit the patient. If he were mentally incompetent, could his wife volunteer him for such research?

Case Study 82. Juvenile Onset Diabetes

There have been two families counseled for the problem of juvenile onset diabetes. Counseling had to take into consideration recent attempts to use HLA linkage as a predicter for such families. Current research indicates that JOD is a combination of environmental (suspected Coxsackie B virus) and genetic factors in about a 50/50 ratio. Some research attempted to demonstrate HLA linkage, with HLA-D,[53] and to chart the likelihood that a sib will develop JOD (including the possibility of prenatal testing). The physicians felt, however, that application for counseling was guarded and that application for prenatal testing was not to be recommended. Newer analysis and research[54] calls these figures into question. Should this new information be used with JOD families? Should they be asked to participate in HLA typing research? Can they volunteer their children for such research? Does the questionable status of the theory have a bearing on the decision?

Case Study 83. Muscular Dystrophies (MD)

A hospital has had frequent counseling involving certain types of dystrophies, and has an ongoing research program involving secretor-gene linkage as a possible prenatal test.[55] The test in-

[53]P. Rubenstein, N. Suciu-Foca, and J. F. Nicholson, *New England Journal of Medicine,* **297,** 1036 (1977).

[54]R. S. Spielman, "Susceptibility to Juvenile Diabetes," Genetics Regulation Series, University of Rochester, March 15, 1979.

[55]Doris Teichler-Zallen and Richard Doherty, "Amniotic Fluid Secretor Typing: Validation for Use in Prenatal Prediction of Myotonic Dystrophy," *Clinical Genetics* **18,** (1980).

volves determining the secretor status of individuals: secretors' body fluids contain soluble antigens A or B, or for blood type O individuals, soluble H-antigen. Saliva is an excellent source. The secretor gene is linked to the myotonic dystrophy gene, and if the parents' and grandparents' linkage associations are known and if they are appropriate in the parents, MD can be predicted with about 85% accuracy in an infant. The research involves blood and saliva sampling, very non-invasive procedures, and is being done with volunteers who had been patients of the program, but who have had no family history or problem with MD. Most had come because of maternal age for amniocentesis, although some had Tay-Sachs, spina bifida, and other genetic problems. Very young babies participated in this research, with the consent of their parents. These were healthy children with no involvement with MD. Did their parents have an ethical basis for volunteering them and did their involvement with the disease make a significant difference in considerations of research justification?

I want to consider what the ethical base for surrogates really is, and this cannot be adequately done under a rights model. It is true that we often justify this as a guardianship or a rights protector approach, but how do we justify assigning someone else as a guardian or protector under this model? In realistic social terms, this is easy enough to see. Courts can step in through legal decisions, or we can justify protection on the basis of *quid pro quo* existing social agreements. But this is not what a rights model in ethics is about. It is not an empirical social role model, based on a working social arrangement that can change, that can give (and therefore withdraw) rights. It purports to justify certain ethical generalizations that will hold for all societies for all times, to ground certain rights that cannot be given or taken away. Social roles cannot serve as such a ground. Now Levi-Strauss' structuralism may be right, so that at this point in the evolutionary process it may be possible to identify certain deep structures that generate possible role combinations, but this deep structure is not a static concept. Since Levi-Strauss finally embeds his deep structures in the biosphere (and not the other way around),[56] and since the biosphere is in process, an ethic based on social roles is very different from what rights theorists, including Rawls,[57] have in mind.

The other model, that of natural rights based on natural law which is in turn based on a supernatural law-giver, will work only

[56]Jean-Marie Benoist, *Op. Cit.*, pp. 26–36.
[57]John Rawls, *A Theory of Justice*, Harvard University Press, Cambridge, Mass., 1971.

if the theological foundation will work, and for every philosophical effort to construct a rational justification for this foundation,
there is an equally impressive philosophical counter-argument in
the tradition.

There is no way I can see, then, under a rights model, to legiti-
mize a transfer of rights from one individual to another. Either
guardianship depends on legal models that are not universal or
immutable, or it depends on social bonds that also can vary. I am
my "brother's keeper" for a number of social and empathic reasons, but not because there has been a mysterious transfer of
"rights" to my person. We act as if this made sense because we
confuse the two concepts of "rights" and use the legal and social
role argument when we should instead be working with the universal and immutable concept.

Will a utilitarian model serve any better? The surrogate would
step in in order to increase the good for the greatest number, but in
this case actually anyone could be a surrogate at any time, if
justified in this way. Further, the surrogate would hardly be a protector of the child or the incompetent, but rather of the general
good, and this is not really the sense of surrogate that we are after.
The surrogate would not really be a surrogate for the individual,
because the surrogate's function would not replicate a mature and
competent individual's function, but rather would give us a social
cost/benefit function. This may be a useful role, but it is not a surrogate's role as we understand it. It also would allow a great deal
of nonconsenting individual sacrifice for general social goals,
which is what the surrogate's role is supposed to guard against.
Children and incompetents would continue to have no input into
decisions concerning them, and under this model no one would
function "as if" they represented their needs and interests, which
is what we mean by a surrogate. Such individuals could always be
treated solely as means, if macro-ethics justified it.

Is there a solution, then, in terms of respect for persons, in
terms of an imperative to treat persons not solely as means (objects), but also as ends? Again, the Kantian ethic depends on a
view of humans as embodying reason (and nothing else importantly) and doing nothing to violate the principle of noncontradiction. Why we should accept this, when our primary processes obviously do not and when we are not even quite sure what we mean
by identity, let alone noncontradiction,[58] I am not clear on. But

[58]Hector Neri-Casteneda, "Identity and Sameness," Philosophia 5,
(1-2), 121 (1975).

suppose we are in agreement on this, that we will do nothing to ourselves or others that would involve a contradiction, since we are part of a rational kingdom of ends. Well, for one thing, we would not need surrogates, since we would all consider behavior involving another human being in the "as if" manner, as if they were also an end. That is, of course, the lure of Kantian ethics. Let me choose a specific example now. We are asked whether to allow a mentally incompetent individual to participate in such a high-risk experiment that it could be described as self-destructive or suicidal behavior. Can we allow it if this maxim were to be a universal law of nature. Kant assumed we could not because it would involve a contradiction, the ultimate base for ethics. He thought he demonstrated the contradiction for self-destruction or suicide in the following manner: I will shorten my life or do not care whether my life is shortened because it holds no value for me (there is more evil involved in living than good, so because I love myself I will kill myself). Kant equated self-love with the function of prolonging and preserving life; hence, to love oneself meant preserving oneself, in contradiction to destroying oneself. Contradictions, however, are sometimes ephemeral things, appearing or disappearing depending on interpretations. We have a logical scheme for contradictions; when we try to plug real events into that scheme, we lose our certainty and sometimes our contradiction. Kant's contradiction depends on interpreting self-love as self-preservation, but these are hardly identical, nor does one necessarily imply the other. We can preserve ourselves because we fear the process of dying and yet loathe ourselves. We can love what we are so much that we will refuse to accept a change or a deteriorative process and preserve ourselves at some point by ending future possibilities that would destroy what we value (by destroying our future projects, in other words).

What I am saying is that contradictions are generated unproblematically only in formal logic and that in the empirical world we can usually interpret them away. We certainly can in this case. (I don't want to try to argue now that we can do so in every case, and therefore that any maxim can be a universal law of nature, but I hold out the potential for such an argument.) We could, without contradiction, allow a mentally incompetent individual to participate in a suicidal experiment. Out of compassion we would not; not out of a principle of noncontradiction. Therefore, not only would not the surrogate have a place in any respect-for-persons models, but everything else would, in spite of the lulling nature of words like "respect."

What is the justification for surrogates, then? The nonrational elements of social bonding ground it: love, concern, empathy, shared experiences, assumed responsibilites. It is a role based first on emotive considerations of the other's needs and interests. Again, because we are social creatures, we place large parts of our selves in the hands of others, most noticeably in childhood when changing or withdrawing that relationship is beyond our control. At other times in life, when we are very sick, unconscious, incompetent, the pattern repeats and significant others make the choices they think we would make, on the basis of experiences shared with us. When social bonding has been inadequate, or when the significant others are themselves incompetent, there is no really ideal solution. Social roles and legal protection offer some relief, but they are not a totally satisfactory substitute.

Finally, let me turn to the question of levels of intervention or invasiveness in research.

Blood Sampling Studies, Amniotic Fluid Sampling Studies, and Psychological Studies

Case Study 84. Fetal Cell Sorter

A tap for amniotic fluid is a fairly invasive procedure, and since it is optimally done no sooner than 16 weeks gestation, any selective abortion decided on becomes a major medical intervention. Results being equal, physicians prefer the least invasive procedure and the least amount of intervention. In view of this, research is being conducted on the possibility of using maternal blood sampling as an alternative to some amniotic taps. Fetal blood cells circulate in the maternal system, so if a way could be devised to sort out fetal blood cells from maternal blood cells, we could collect the required fetal cells for culturing from a maternal blood sample rather than a tap. Using laser and computer technology, along with fluorescing stains, such a procedure is being researched. There may be some interesting implications for this. If testing can be so simplified, it could become available to many more people, as far as ease of collecting cells for culture. However, culturing and analysis by the lab would still be as labor and time intensive, so either facilities would have to be expanded or restrictions on accepting patients (in terms of age, need, etc.) would have to be implemented or new lab techniques using the machine itself would have to be developed. If the time frame for testing could be moved up, selective abortion could be done at an early, less traumatic period (before the fetus' movements were felt, for example). Would this make the abortion decision too easy in the sense of reducing sensitivity to what was involved? If the test could be done quite early, routine sex selection through selective abortion might become a possibility. Participation in such research would be quite low risk for the individual. Should participation in such research also involve questions about the social risk/benefits of such research? Would issues of over-medicalization or control of the application of the technology be appropriate considerations for potential subjects?

Case Study 85. Spina Bifida and Alpha-Fetoprotein Research

A few years ago, the alpha-fetoprotein test for neural tube defects was still experimental and our center was sending such lab work to a laboratory on the east coast doing the research. There was concern with the possibility of false positives and confidence in ultrasound confirmation was not that high. One couple's test results represented a problem. The test was borderline; however, the maternal serum test was negative and two ultrasound scans were also negative. The researchers recommended no further invasive procedures such as amniography. Three experts were directive about continuing the pregnancy and limiting any further intervention. Now the test is more routine and done on every tap rather than only those with indications for NTD. Our lab, and others, handle the analysis. Ultrasound has become refined enough to confirm or disconfirm the AFP test and what was an experimental procedure has now become a standard procedure. The first subjects of such procedures assume more risk, of course, until experience and refining techniques and theories finally place the procedure into routine risk levels (not no-risk levels, however). The experts remain concerned with levels of invasiveness and intervention in physical procedures, and in that role frequently limit the risk-taking that their subjects may assume. As I discussed in Chapter 3, autonomy is not the only value to be considered in risk-taking.

Case Study 86. Anxiety Study

A young couple, both lawyers, want only one child, a healthy baby. They wanted all possible testing. During the session, they were asked if they would like to take part in an anxiety study, and the wife agreed. Another couple, both clinical psychologists, came for maternal age testing. The wife felt the counselor's information was unnecessary, repetitious, and wanted to skip over it. They do not perceive a high risk, but wanted amniocentesis. They were asked, and agreed to participate in the study. The study is designed to produce information that might be helpful in reducing anxiety concerning actual test procedures. The research may or may not directly benefit the subjects. The experimental procedures are not at all physically invasive (questionnaires, and videotapes, for example), but could be mildly invasive psychologically (in one very broad sense of the word invasive). Should there be different levels of concern for physical and psychological research intervention? Should all program patients be asked to participate in the study or should the highly stressed ones be screened out? Is there a difference in perception of how necessary the research is between physiological and psychological research, and is this an accurate perception?

These three case studies illustrate divisions we could make in considering degrees of invasiveness of intervention in research (or for that matter, in general medical intervention). The first possibility, of course, is the split between physiological and psychological invasive procedures. There are some reasons for this apparent dichotomy, the most important being our assumption that the results of physiological intervention are much more apparent, readily at hand, quantifiable, easier to manipulate than the consequences of psychological intervention. We have a better handle on the causal web (not linear chain, please, because then our handle is a pseudo-switch). This actually says nothing about the real invasiveness of either intervention, however, only that we are more aware of the results of one than the other. In terms of total body-system functioning, in fact, a psychological breakdown may be more apparent than some physiological breakdown. It is on the reduction level, because we have narrowed our interests and purposes and acted "as if" only a small segment of the web needed to be considered, that our results seem so apparent, so self-evident. If we expand our view to larger and larger subsystems, the apprehension of consequences can often take on the same tentativeness that the psychological procedure can.

In fact, the same analysis also applies to the psychological realm. Thus, if we narrow our interest to behaviors that are relatively simple, under controlled environmental conditions, the results of a psychologically invasive procedure are equally as apparent and at hand for observation and analysis. Both procedures can be highly invasive and either have readily apparent or non-apparent consequences. At the same time, we sometimes feel that physiological intervention can be minor and easily controlled, whereas psychological intervention is always major and problematic. This stems from the old mind/body dualism, and will explain an aside I made in the last chapter. When we divorce mind and body, or human action, from event, we invest (implicitly, unfortunately) our self and integrity in the mind half, so that assaults can be done to the body without jeopardizing what we think of as "us," our selves, our integration, our control. We view ourselves as immune from most physically invasive procedures, which we imagine as something done to our bodies, but not our selves. However, *any* manipulation of our minds, then, becomes a highly invasive procedure, an assult against our integrity, self-dignity, and control. There is, even in the interventionist science of medicine, a reluctance to intervene psychologically. This also creates the uneasy attitude toward psychotherapy that exists not only in the general culture, but in medicine as well.

Empirically and conceptually the mind/body dualism is an artifact of time-lag. The fact that "after-death" experiences could be seriously offered as evidence *without* determining three flat electroencephalograms during their occurrence is evidence of that. The state of the central nervous system (the state of the body) was simply assumed to be implicitly unimportant, *which was the very thing the experience was supposed to demonstrate.* This again is a case of circular reasoning. Time-lag in a social system is not an easy thing to prevail against. Bioamine research also demonstrates that the dualism is empirically inaccurate,[59] but has had little effect on our implicit humanistic paradigms. Finally, systems theory (especially hierarchy levels) can meet all the objections of those who support such a dualism by arguing from their straw man of oversimplified reductionism. Systems theory is a viable, heuristic model, that is usually either ignored or misinterpreted because of the older paradigms.

I can make the reasonable conclusion, then, that both physiological and psychological procedures can be highly invasive or non-invasive, and that my discussion of the ethics of research

[59]J. D. Barchas, H. Akil, G. R. Elliott, R. B. Holman, and S. J. Watson, "Behavioral Neurochemistry: Neuroregulators and Behavioral States," *Science* **200**, 964 (1978). Philip A. Berger, "Medical Treatment of Mental Illness," *Science* **200**, 974 (1978); Burr S. Eichelman, "The Aggressive Monoamines," *Biological Psychiatry* **6**, (No. 2), 143 (1973).

applies equally to both research procedures. The only difference between the two types of procedure I can see is that informed consent can sometimes (not always) prove more of a problem with psychological procedures because of risk of influencing or invalidating results. (The placebo effect is a well-known bit of evidence that informing the patient can invalidate physiological research as well. The double blind experiment is constructed for a reason.)

As levels of invasiveness increase, changes occur in terms of the balancing of patients' interests, patients' needs, individual benefits, social benefits, expert decision-making. I have already mentioned that the expert role restricts patients' interests and even needs. Benefits are also seen in the context of risks (highly invasive procedures), but this is also symmetrical. Risks are evaluated in the context of both individual and social benefits. Whether there is some cut-off point for levels of invasiveness in research is debatable. There is a history of highly invasive self-experimentation among researchers that, in the current period when collaborative research is nearly required, may be nearing an end. Whether to view this behavior as self-destructive or humanly grand is open to interpretation, but I tend to think it was more akin to the Grand Prix racer than the bridge abutment fatality. Whether to engage in this kind of activity with volunteer patients is more questionable. I think realistically that terminal patients are more willing (sometimes strongly desire) to take high risks, but that very willingness and desperate situation imposes a responsibility on the expert to at least have some animal data and/or theory grounding the gamble. A useless, meaningless risk for a terminaly ill patient remains a useless, meaningless risk that unnecessarily uses up part of the time he has left.

In addition, we are under no obligation to agree to allow self-destructive or self-punishing individuals to participate in research, if it is clearly determined that this is the case. The only caveat is that it also makes no sense routinely to impose a paternalism that would deprive marginal individuals of chances for self-determination.

This discussion has earlier demonstrated little difference in justification between therapeutic and nontherapeutic research. Some difference emerges as the level of invasiveness rises, because the benefits are higher for the individual in therapeutic research, and this is directly related to levels of intervention in terms of risk/benefit.

Finally, I have raised the issue of looking not only at individual risk, but also at social risk in deciding whether participation in research is justified. This is too complex an issue to handle adequately in this chapter. Let me just make some very brief observations. The past example of the recombinant DNA research debate is not an example I would choose to illustrate the proper analysis of social risk, unless in a negative way. We have to do at least adequate projections, a respectable futurism, to arrive at a reasonable decision. The ethics of intervention stresses adaptation to a continuously changing environment, so high risk should not be equated with change in making our evaluation. And last, systems maintenance (stasis) is not a steady-state, but an oscillating, condition, which means that the assumption of risk is built in.

In view of all this, and to give us a perspective on research, I want to consider the large, naturally occurring experiment in which we are all subjects. I could say, actually, that the evolutionary process was such an experiment, which of course it is, but I would like to narrow that a bit. I want to consider a few case studies illustrating our participation in an experiment without good design or control groups: the introduction into the environment of a large number of substances not naturally occurring at a significant rate.

Future Effects
of Environmental Contaminants

Case Study 87. Alcohol Abuse, Librium

A woman 2-months pregnant was referred by a mental health center. She was scheduled for a therapeutic abortion because of heavy alcohol consumption and use of librium during treatment. Alcohol is a socially available teratogen, and librium and valium are among the most commonly prescribed drugs in general practice, two psychoactive substances introduced into our pharmacopeia. The mental health center had thought amniocentesis might be of help for this patient (it would not be), but the genetic counselor really could give no risk information for librium, in terms of mutagenic or teratogenic effects. Valium, for example, is now suspect in this regard, but both substances are part of a natural experiment anyone taking them participates in.

Case Study 88. Dioxin

This young couple have one healthy child, followed by three spontaneous abortions. The child was born before the couple, very much involved with natural foods and healthy life-styles, moved into a log cabin they built from a kit. The logs were treated with a preservative that may contain dioxin. The three abortions occured after moving into the log cabin. The physician will do chromosome analysis to rule out chromosomal abnormalities as the cause. The possible dioxin contamination remains a question mark. The introduction of dioxin into the environment has been an uncontrolled, unanalyzed experiment.

Case Study 89. Chemical Dyes

A couple was counseled whose baby died within an hour of birth from multiple abnormalities: lack of eyes or kidneys, presence of finger webbing, scoliosis. The baby's chromosome analysis was negative. The wife was worried because she had industrial exposure to acetone and chemical dyes, and her request during pregnancy to be transferred to another area had been turned down. In her distant family history, there is a case of absence of all extremities, as well as a case of a single kidney. Again, this undesigned experiment with chemical dyes can lead to no scientific conclusions.

Case Study 90. Whole Body Radiation

A nurse, now pregnant, wanted counseling because of concern over whole body radiation she had received in childhood for acne. Although amniocentesis is limited in the information it could give her, some chromosomal breakage or linkage might show up. She had had 3 months of radiation treatment. She elected to have the test, although her counselor assured her the risk was probably minimal. The introduction of therapies is also an experimental situation and was done without careful research design.

Case Study 91. Phosgene and Benzene

This young couple have two healthy children and then had three spontaneous abortions. The husband had been exposed to high levels of phosgene and benzene. A chromosome analysis will be done to eliminate that possibility, but again, little can be said about the industrial exposure.

Case Study 92. Radiation

A child diagnosed with Turner's syndrome died within a few hours after birth. The family history was negative, but both parents had had extensive radiation exposure. The physician stressed that the recurrence risk was less than 1% and that prenatal diagnosis could be done, but played down possible radiation effect because he did not want to induce parental guilt. Again, because of the lack of design of most natural experiments, little can be concluded from the radiation exposure.

These case studies are frustrating, and we all share the frustration. Whether introduced for benefits (prednisone, dilantin, X-ray, DES), or as an industrial byproduct or discharge, substances in our environment have made us all research subjects, often without consent. In part, this is a necessary trade-off that accompanies the benefits accorded by industrialization, and ultimately civilization.

The unethical aspect that is particularly frustrating, however, is the uselessness of most natural experiments. Because of lack of design and inability to retrieve information, the risks are purposeless, meaningless. Time is spent arguing the fine points of scientific proof requirements, and a lack of financing for primate or longitudinal studies compounds the difficulty. The ethics of natural experimentation, which should concern us all as subjects, has hardly seen the light of day.

Summary

I think the discussion above has demonstrated little difference in justification between therapeutic and nontherapeutic research. Some difference emerges as the level of invasiveness rises, because the benefits are higher for the individual in therapeutic research and this is directly related to levels of intervention in terms of a risk/benefit analysis. But the real base for research requires the recognition of the affective components necessary for an ethical system: compassion, attitudes toward pain and suffering, and sociability.

Again, when we consider the requirements of the surrogate role, and attempt to ethically justify that role, the traditional ethical systems fail. Rather, our use of surrogates is grounded again in affective components: love, concern, empathy, social bonding, sharing of experiences. To make sense of the major elements of medical ethics, then, we must construct an ethical system that places these factors in a strategic role.

Chapter 6

Selective Abortion

Range of Choices and Time Frames

We have finally come round to the ethical issue of abortion, and because the ethical context by which we can make some sense of the literature has now been developed, we are in a much better position to examine this hard question. There are two principal issues to get clear about when discussing abortion:

(1) There must be a clear analysis of killing, specifically homicide, its connection with human worth, and the question of special status for human beings in relation to the rest of the biosphere.

(2) There must be a clear analysis of the different levels of interest and complexity involved in the abortion issue. These levels generate, not new ontological entities (e.g., forests) different from those on the lowest level (e.g., trees), but more complex aspects of combined entities, with emergent characteristics to be pragmatically handled "as if" we were dealing with new entities. The entities are functionally new and functionally significantly differ from those on the reductive level.

Using my proposed systems analysis of both these issues, I think the abortion question can be conceptually cleared up, at least in terms of a rational analysis. I do not believe the issue will be decided by such reasonable means at this time, however, because I feel there are too many layers of culturally programmed emotional responses to it, as well as ways of perceiving ourselves as "different" from the rest of the universe that have been reinforced by the humanities. Nevertheless, an attempt to clear away the conceptual muddle is worthwhile as a prelude to unraveling the sociological and psychological confusions.

Let me sketch the proposed argument in tentative form:

(1) Killing is:

(a) Not an immediately apparent bad action or wrong, because it does not always cause pain, can have good consequences, and is not always felt to be wrong.

(b) Sometimes the best alternative or worth the cost, and therefore an instrumental good.

(2) Human beings are special to each other, but not objectively differentiable from the rest of the biosphere or the rest of the universe.

(3) What has been said about killing in general therefore applies to a subset of killing that differs only in terms of human purposes and interests: homicide.

(4) In terms of human interests and purposes, we frequently make decisions either directly or indirectly to commit homicide, and these decisions are necessary for our very social existence.

(5) Homicide is not a contradiction of the worth of human interests (the worth of human beings and their projects) because killing does not entail a conclusion that the killed has no worth.

(6) Our human interests have led us to make valuations about homicide that distinguish between ethically justified homicide, ethically unjustified homicide, ethically justified murder, and ethically unjustified murder.

(7) Therefore, it is necessary to determine:

(a) Whether abortion is killing, homicide, or murder.

(b) Whether it can be ethically justified.

(8) In order to determine these things, we need a workable definition of what we mean by a human being in actual situations. This is best approached by employing a systems model to examine levels of interest, analysis, or organization.

If we are concerned with killing, we are talking about disrupting a homeostatic system composed of biotic material capable of being partially described on a reductive biochemical/biophysical level. If we were at that more fundamental level, we would speak of "destroying" material, rather than "killing" it. "Killing" is a term already signifying that a more complex level of organizaation is being dealt with, one where stasis mechanisms are operating. A butterfly, a bacterium, a young gazelle, a *homo sapiens* can be killed. A fetus can be killed since it is a stasis system, not by virtue of its potential to become *homo sapiens*. A fertilized egg can be killed since it is a stasis system, again not because of any potentiality to develop into *homo sapiens*.

"Homicide" is a term indicating that we are interested in the taxonomic characteristics of our stasis system: are we dealing with an entity belonging to *homo sapiens*; are we killing a member of the human species? I do not want this term confused with the legal definition of homicide, which depends on an implied notion of legal personhood. In legal homicide, a "person" is killed, someone who by birth has the status of "person" under the law, with very few exceptions in legal precedents. Restricting the term here to the physiological level will eliminate the risk of seeming to employ it in the analysis of two very different levels of organization.

Now, we have to come to grips with "potentiality." We can fairly easily determine which stasis systems are existing members of our species using a number of indicators, one of which is chromosome analysis (even when something has gone wrong with the process and there is chromosome abnormality). Extending these indicators back in time is also no problem at first. When we have reached back to the zygote stage, we can still say that chromosome analysis alone indicates *homo sapiens*, at least *homo sapiens* in process, the other taxonomic indicators now only being potentials. "Potential" is tricky, however, since chromosome analysis will also show that a human egg and sperm cell, given the proper environment (proximity to each other, proper vaginal acidity, etc.) are also "potential" *homo sapiens*. (There are cultural consequences of this sort of thinking: sperm should not be wasted or killed by being spilled on the ground, rather than in a vagina where one of them may go on to express the "potential" for becoming *homo sapiens*, for example. Criticism of or sanction against birth control methods that suppress the expulsion of an egg, depriving it of its "potential" for becoming *homo sapiens*, is another.)

It really is no wonder that appeal to the potentiality principle is an appeal to confusion. We are not clear on what we mean by "fertilized egg is identical to fertilized egg" (A=A), let alone "fertilized egg is identical to person" (A=A'), so the confusion is understandable. "Potentiality" expresses the notion that experience is in process, and that as part of such a process, none of us is a static entity and our interrelationships are complex. But existence and future projection are not equivalent. Processes can be modified or halted, so that our interest in categorizing experience is heuristically served by constructing these categories around relatively temporally stable experiences. We do not count our chickens before they are hatched because A (egg) really is not equivalent to A' (chicken). To give a term that implies characteristics of a *fulfilled* possibility to a potentiality is conceptually misleading and will pragmatically have conceptual deficit rather than conceptual value. The egg, sperm, zygote, and fetus are only projected or future *homo sapiens*, in a more strict sense of that term, and it is therefore not really so clear that we are committing homicide when we kill them. It depends on whether we decide it is necessary to use *homo sapiens* as a term to characterize the uterine and gonadal phases, and such an evaluation is of course always made in terms of human purposes, interests, needs. Our cognitive categories also represent choices, but not random or equal choices. Let me suppose that it would be useful to so extend the category *homo sapiens*, however, to make the strongest case. I will call killing an egg, sperm, zygote, or fetus "homicide" in the sense that I have been using here, and overlook all the philosophic problems with potentiality.

Homicide, in terms of this argument, is not an apparent evil but is, in some situations, ethically justified. Abortion is, in these terms, killing and homicide, remembering the special sense of these terms. Is it murder? "Murder" is a term incorporating a social level of complexity, since we murder "persons." If we wish to use this term in the same way we use "homicide," we inevitably reach the unworkable conclusion that a sperm is a person. The reasoning here is precisely the same reasoning that judges a zygote to be a person, and those who hold certain ethical positions on abortion have in fact concluded that a zygote or a fetus is a person.[60] They accomplish this by confusing levels of interest and organization. "Person" is a term that signifies we are talking about possibilities of social interaction, of social configurations. The maternal bonding interaction is probably the first expression of this. An identifiable and distinct individual (the baby) can interact with another identifiable and distinct individual (the mother), and the process of socialization has begun. There is no possibility for this

[60]See, for example, John T. Noonan, "How to Argue About Abortion," *Contemporary Issues in Bioethics*, pp. 210–17; *The Morality of Abortion*, Harvard Un. Press, Cambridge, Mass., 1970; Sissela Bok, "Ethical Problems of Abortion," *Hastings Center Studies*. **2**, No. 1, Jan., 1974; Gordon C. Zahn, "A Religious Pacifist Looks at Abortion," *Commonweal*, May 28, 1971, pp. 279–282.

kind of interaction *in utero*. The fetus cannot identify and distinguish the mother, only the uterine environment at most. The mother cannot visualize, touch, or feel proximity to the fetus, since the fetus is still part of the mother, not distinct, and not identifiable by her. This social interaction becomes a possibility after birth, and the term "person" in our legal sense reflects this. (It may be objected that fetoscopy can allow a mother to see the fetus, but this is a limited technology that at most could allow a small beginning for the mother toward a social interaction, but no possibilities for the fetus. A social interaction requires participation of both. The fetus physically interacts with part of the mother's internal system *in utero*, but this is not the social interaction, the socialization process, whose possibility is central to our use of "person," a term on the social level of interest and organization.)

If we murder "persons" by definition, then we cannot murder fetuses, since fetus entities are not person entities, and murder makes no sense on the fetal level, only on the social level. Does such a social definition or use of "person" create problems? For example, is a feral child a person? Is a vegetative individual a person? Will this "systems analysis" allow these sorts of killing as well as abortion? The answer is that it does not automatically preclude it in some instances, and whether such homicide can be ethically justified would depend on the interests and needs of another person to the transaction. Let me say that the cases of feral children have not been cases of completely asocialized children. It may be impossible for a human child, or the young of other social species, to survive nonsocialization even physically. Studies on the rhesus monkey model for depression indicate that there is a withdrawal response that will eventually be fatal.[61] So I suspect that the example of a feral child would be contrary-to-fact, and I have philosophic prejudices against taking contrary-to-fact examples seriously, stemming from empiricist and pragmatist predilections. The vegetative individual, whether newborn or very old, is another matter, to be taken quite seriously. The newborn who would be so defective as to be incapable of *any* social interaction and yet remain viable would not be common. Such an individual would never have been a "person" and never could be. The individual who, because of trauma or health deterioration, has become vegetative enjoys a past history of personhood, although that status no

[61]H. F. Harlow and M. K. Harlow, "Social Deprivation in Monkeys," *Scientific American* **207**, 136 (1962); S. J. Suomi, H. F. Harlow, and W. T. McKinney, Jr., "Monkey Psychiatrists," *American Journal of Psychiatry* **128** (No. 8), 927 (1972).

longer has reality. (Again, I am not talking in the legal mode. Both of these individuals are legally persons. Nor am I talking about static terms, conferred on an entity once and for all, but of a changing experience, in process.) Ethically, choosing actions regarding these individuals involves the interests and needs of persons involved in their situation as the primary concern. Memories of past social interactions will tend to result in involved persons treating the once-interacting individual as if it were still a person, and there is usually a strong need to do this. That need is the important factor, however, not the pseudo-personhood. Whether homicide would be ethically justified or not depends on the needs and interests of those concerned (and the long-term consequences, effects of the decision on the decision-maker, etc.), rather than a sanction against killing persons.

As a matter of fact, using the term "murder" for the person level, it is not immediately apparent that all murders are ethically unjustified. We indirectly kill "persons" routinely (raising the speed limit to 65 mph, not converting chlorination water treatment plants to carbon filtration systems, switching from gas-burning to coal-burning systems, congressional authorization for war), and it is not apparent that these are all unjustified. We kill "persons" directly (self-defense when physically attacked, the executioner's role in capital punishment, shooting an intruder). The indirection or directness of the method and motive should hardly make the ethical difference between good and evil on its own weight. Directness may say something about responsibility in regard to purposes such as punishment, but surely, if the consequence is bad, then the action (killing) was evil, whether there was a complicated causal web or confused and contradictory intentions. This is the problem with the double effect argument.[62] Whatever the means and intention, if one assumes the consequences are bad (the death of the fetus), the indirection and motivation will not change that, and it would be very peculiar, in ethics, to ignore the consequences. In fact, I hope I have successfully argued that all ethics makes sense only as consequentialism, that we choose a rule that guarantees more frequent good results, an intuition whose following we believe produces an on-balance good universal order, or a situational calculus for good consequences.

[62]Philippa Foot, "The Problem of Abortion and the Doctrine of the Double Effect," *Moral Problems*, James Rachels, ed., Harper & Row, New York, 1971, pp. 29–41.

Thus although abortion is not conceptually murder (killing on a person level), if it were, it would not be an immediately apparent evil.

My definition of "person" has been made in terms of systems theory, which is actually a significant advantage here. Others have introduced the concept of "person" in order to make some sense of the abortion problem and to humanely attempt to avoid the large increase in pain and suffering that categorizing egg, sperm, zygote, embryo, and fetus in the very same way as functioning social individuals necessitates.[63] But they have all been generally dissatisifed with such a concept of personhood because there was little theoretical justification for the different categories, and because the term "person" thus became quite free floating, and seemed too confused itself to offer much help in clarifying the abortion problem. Systems theory, however, can offer the theoretical underpinnings for the use of such categories, and in fact, demands their use if confusion is not to result. The development of hierarchy theory and the recognition of functional entities places an analysis of "person" on firm ground, both philosophically and scientifically.[64] We can now speak meaningfully of persons and fetuses.

If we have kept our level-of-organization terms and entities unmixed, we can now say that abortion is killing, and not so clearly "homicide," but certainly not murder. If these statements hold, then such killing can be the right action in some situations, and the ethical justification will involve an evaluation of the consequences (including the consequences to the actors). These consequences must be evaluated in terms of the interests of the "persons" involved, including the effects on them and the broader social network. The evaluation also needs to follow the ecological principles I outlined in Chapter 1; that is, that all relevant alterna-

[63]See for example, Jane English, "Abortion and the Concept of a Person," *Contemporary Issues in Bioethics*, pp. 241–243; Mary Anne Warren, "On the Moral and Legal Status of Abortion," *The Monist* **57** (No. 1), January 1973; Sissela Bok, *Op. Cit.*; Bernard Haring, "Theological Evaluation," *The Morality of Abortion*, John T. Noonan, Jr., ed. Harvard University Press, Cambridge, Mass., 1970, pp. 123–145.

[64]H. Soodak and A. Iberall, "Homeokinetics: a Physical Science for Complex Systems," *Science* **201**, 579 (1978); William Wimsatt, "Reductive Explanation: A Functional Account," *Basic Studies in Philosophy of Science*, Vol. 30, Reidel: Dordrecht, 1976, pp. 647–86; Howard Pattee, *Hierarchy Theory: The Challenge of Complex Systems*, Braziller, New York, 1973.

tive actions and inactions and their results must be calculated in reaching the decision.

I have spent a good deal of space trying to develop a framework for this analysis because the abortion issue *is* so confused. Selective abortion in a prenatal detection program adds to the abortion controversy the problem of devaluing human beings with genetic defects, but I believe that this linkage between killing and worth has been successfully shown (in chapter 4) to be unnecessary and conceptually unwarranted.

It is time now to take a look at how the abortion decision is actually faced and made in a working program, and I want to do this for each step of the process to illustrate the options that continually remain open to the patients.

Counseling Decision
After Physician Referral

Not everyone who is referred to a prenatal detection program by a physician decides to take advantage of even the genetic counseling available. Often an appointment is made, and then broken. Referring physicians vary in their directiveness. Some are quite insistent, even making the appointment themselves. Others urge the counseling and testing. Some simply inform patients of its availability, and there are a few who make referrals only after their patients have actively requested it.

Case Study 93. Maternal Age

The wife is 37 and pregnant, with a history of spontaneous abortions. She has one child in good health. She had decided to take advantage of counseling and consider amniocentesis. Although she was over 35 for her last pregnancy, her physician had not informed her of the availability of the test. This could have had serious legal repercussions for the physician if the pregnancy had resulted in a Down's child.

Case Study 94. Maternal Age

This was a planned pregnancy. The couple had read about the test, and both they and their referral physician were mutually agreed on counseling and testing. This was a well-informed couple.

Case Study 95. Maternal Age

This is the 40-year-old wife's seventh pregnancy. All her children are healthy. Her obstetrician referred her, but she had little knowledge of the reason for referral or about prenatal testing. She was primarily following her physician's instructions.

Case Study 96. Maternal Age, Down's History

The couple have tried for eight years to become pregnant and were finally successful with cycling hormone. The father was very negative and hostile to the interview. He thought the referring obstetrician was "crazy" and that this was basically a woman's problem. The counselor did manage to explain the situation to him, and he left feeling more positive about the obstetrician's referral.

Case Study 97. Maternal Age

The wife is 38, and this is her fifth pregnancy. She brought up the advisability of amniocentesis with her obstetrician, who then gave her incorrect risk figures (too low) for Down's syndrome. She has no concerns about other birth defects and is ambivalent about this pregnancy.

   The referring physician can be directive or nondirective about giving information concerning the availability of prenatal testing. However, for women 35 or older, there is a legal obligation to inform about amniocentesis. The physician can still choose to be negatively directive about the test, but needs to tell patients that such a test is available to them. This informing itself requires the patient to make a decision about seeing the genetic counselor or not, and opens up a net of options and decision-making. The patient does not have autonomy as far as opening up these options is concerned. Legal and medical considerations will determine whether information about the test is conveyed, and the case studies illustrate that the physician can make both informing and non-informing decisions in the role of expert. In maternal age cases, the patient does not have a "right" not to know or to practice denial; the patient will have to make a decision about counseling. It would be impossible by definition to give the patient an informed choice about being informed. The expert makes the first decision (or the circumstantial social availability of information), and it is this that allows expanded options. There is no doubt that this process can be uncomfortable, but as I discussed in the section on anxiety, the mechanism is part of the human condition. Thus, if we value functioning in some adequate relationship with our changing environment (what we really mean by developing our human potential), we incorporate anxiety or possibilities of discomfort into that choice.

Prenatal Testing Decisions

Once informed of the testing availabilities and counseled about them, the patient must decide whether to make use of this option, and now becomes a major figure in the medical interaction.

Case Study 98. Maternal Age

Maternal age is one of the major counseling categories. In this case, the wife is 39, and had already made up her mind before the counseling session to have amniocentesis done. She was totally accepting of her physician's recommendation that she have the test. Although a member of a religion that opposes abortion, she had also already made up her mind to abort if a serious defect were found.

Case Study 99. Maternal Age

This woman was 40, had been referred by her physician, and was very confused about the test option and the need to make a decision now confronting her. Ethical concerns about abortion were not the reason, since she had no qualms about aborting. She actually seemed apathetic and it was not a satisfactory interview. Anxiety was not apparent. Instead, she seemed rather dazed by the option and left without making a choice.

Case Study 100. Maternal Age, Medications

This woman had initially decided against having amniocentesis. Although feeling abortion should be an individual choice and open to anyone, she could not approve of it for her own case. Then she became ill and took medication, which sufficiently concerned her that she changed her mind and decided to have the test. However, the counselor told her the test would not be able to pick up possible damage from the medication, which because of her psychiatric history had been Elavil. Decision-making again became difficult and she wanted the counselor to tell her whether to have the test or not. The physician was nondirective and would not make the choice for her.

Case Study 101. Maternal Age

The couple had not been aware of the test until their physician told them. The wife is 37, and they had also been unaware of Down's syndrome or the age risk. They will now have amniocentesis because their physician told them to have the test. If the test were to show a serious abnormality, their decision to abort or continue the pregnancy would also be based on what their physician said.

Case Study 102. Maternal Age

Of a medical group of obstetricians, one mentioned the test to the patient. When she came, she had not decided whether to have it, and was very upset when the abortion option was mentioned (she cried). However, after being assured that no precommitment to abortion was necessary and that she could use the test information to prepare for whatever problem might be indicated, she had the test scheduled for that afternoon.

Case Study 103. Spontaneous Abortions

These abortions did not fit a chromosome abnormality pattern. Two had occurred at five and six months and the deaths owed to prematurity rather than any discoverable defect. Although this doesn't fit a translocation pattern, chromosome analysis was offered as an option. The couple decided against the test based on the low odds and cost (actually, with insurance coverage, cost was probably a minimal reason).

Case Study 104. Encephalocele

The husband was opposed to the test because of the small risk factor (US and Canadian studies indicated no significant risk factor). Also, he is opposed to abortion, which he views as homicide. The wife would have liked to have the test, but the decision was made not to have it.

These case studies illustrate the range of decision-making encountered in day-to-day practice and the problems some patients have with it. There are two major issues here: (1) How directive should the counselor be in assisting the patient with a decision or making it for the patient, and (2) Should a precommitment to abortion be required. I talked about precommitment requirements in Chapter 1. Ideally, at this step in the option net, abortion should be discussed as one of the options, so that the patient understands the situation; but it is not necessary to link the decision to undergo amniocentesis with the decision to have an abortion if in fact the test indicates a serious defect. The test decision can be made conceptually before an abortion decision is made, and need not automatically lead to such a decision. As always, however, there are exceptions. In amniocentesis, the risk from the procedure is very small. However, in an experimental or highly risky test, the test benefits would have to be sufficient to balance the risk, and this decision, as I discussed previously, could be made unilaterally by the physician in the role of expert. The physician could request that the decision to have a highly risky test be made concomitantly with the decision for abortion, otherwise declining to perform the test on the grounds that it would amount to self-destructive behavior and unjustified medical intervention.

In addition, social choices or constraints may also force a decision. Lab personnel, equipment, and available physicians are all limited resources. If these resources are pressed past their limits, priorities might have to be established. Such a ranking could (and does in Andre Boué's Paris lab, for example) involve age cut-offs, and could also involve abortion precommitment, seriousness of defect, level of risk, and so on. This raises the ethical question of "fairness" in a context of limited resources. Philosophically, priority ranking is one suggested means of meeting the obligations of a fairness principle. It has been criticized as involving subjective choices, however, and making assumptions about what we value, as well as being prone to manipulation and influence. An alternative suggestion has been the "first come, first served" method of limiting (rationing), which also has apparent elements of unfairness. Although it is not subjective, that is precisely its problem. There are many times when human beings want to and feel perfectly capable of making choices or decisions. Thus, in the triage situation, we do not want to treat the individual who will easily survive without treatment ahead of a seriously injured, but salvageable if immediately treated, individual, even if the person who will survive without treatment arrived first. Nor do we want to treat the dying and obviously unsalvageable individual before the treatable one, regardless of the order of their arrival. Leaving the decision to the random order of arrival abdicates our human function of choosing, ignores our human characteristic of interest, purpose, and need, negates our compassionate feelings, and replaces intervention with acceptance. In short, it is not ethics at all, certainly not as I have described and grounded ethics, but rather purposeless reaction to the environment instead of a purposeful interaction that develops our meanings, our humanness, in the context of the functions of the physical world.

The last alternative to human choosing was proposed by Nicholas Rescher.[65] All things being equal, he suggested chance as the means most acceptable to the human sense of fairness. Of course, all things have never been equal, and there never are any duplicate candidates for a limited resource. But assuming we have a pool of candidates whose differences are minimal, he proposed drawing lots to select the final group. This is again abdication of human choice and allowing nonhuman circumstance (in this case, chance) to replace a true ethical decision. It is argued for on the basis that human beings would feel it more fair to be selected against by blind chance (blind fate?) than by their fellows. Actually, mythology is one of the strongest arguments against attributing this feeling to humans. All sorts of elaborate systems have been constructed to avoid concluding that chance is the major operating principle of the universe, and this pattern of system creation is worldwide.[66] The willingness to assume painful guilt rather than accept chance as our explanation for a range of disasters (from genetic to natural) is also a readily observed psychological phenomenon. Human beings do not appear to be comfortable with chance. It removes the feeling of control, of childhood security and omnipotence, which may not be rational, but is certainly operative and cannot be ignored. Thus, as a reason for choosing lottery over decision-making, it is quite inadequate. Nor is it so apparent that chance can be described in terms of "fair." Chance is hardly fair or unfair. "Fair" has meaning only in terms of our human choices, and whether they satisfy us as representing as many needs as possible. If chance is neither eagerly embraced as a way of viewing the universe, nor relevant to the concept of "fair," it has no plausibility as an alternative. If we wish to function ethically, it never had.

Therefore, if we ethically ration our resources, we do it in terms of human choices, human interests and needs. It is conceiv-

[65]Nicholas Rescher, "The Allocation of Exotic Medical Lifesaving Therapy," *Ethics* **79** (No. 3), 173 (1969).

[66]Samuel Noah Kramer, *Mythologies of the Ancient World*, Doubleday, Garden City, New York, 1961.

able that, with such limited resources, we might eventually need to limit amniocentesis to those who would abort a defective fetus. We do not have to do that yet. The decision to abort occurs after the test decision, although the patient may have privately reached a decision to abort before counseling and test results. Such a decision, nevertheless, is subject to change until the actual abortion.

Abortion Decided Affirmatively or Negatively Before Testing

Case Study 105. Down's Syndrome

The program had its first positive test result for Down's syndrome in two years, and a selective abortion was scheduled. The couple had decided they would abort an affected fetus before testing was done. Their decision remained unchanged when the actual choice had to be made.

Case Study 106. Maternal Age, Thalassemia

The wife is 42, pregnant, with a history of two ectopic pregnancies and one spontaneous abortion. The couple has two adopted children. Since they are both of Sicilian background, thalassemia carrier testing as well as amniocentesis was suggested. Before counseling, they were undecided about abortion. After receiving the information, they decided that if there were any problem, they would want to terminate the pregnancy.

Case Study 107. Maternal Age

The wife is a clinical psychologist, is now pregnant, and the couple has a young son. She felt well-enough informed that she did not want the counselor's information and had already decided on the test. Their son was with them and they also wanted him present during the test. They had no doubt they would have an abortion if the test indicated a genetic defect.

Case Study 108. Maternal Age

Although the patient is the wife of a physician, she still has not seen an obstetrician or confirmed the pregnancy. Her husband heard abut the test from a colleague. The couple has two young children, feel they would be adversely affected by a seriously detective sibling, and would therefore abort if amniocentesis indicated a problem.

Case Study 109. Maternal Age

This is an older couple who decided they would not abort a Down's fetus, but wanted the test anyway. They felt the test information would allow them to prepare for a Down's baby and therefore would be of value to them.

Case Study 110. Hemophilia

Two of the wife's brothers have very severe cases of hemophilia. For religious reasons, they decided before testing that they would not abort. Our lab can do carrier testing for the wife and amniocentesis could determine if it were a male fetus. But only more experimental procedures are available for prenatally testing the male fetus. The couple want amniocentesis even if abortion is not one of their options, and the test will be done.

Case Study 111. Spina Bifida, Medication

There are a number of problems in this case. The wife had a 4-month-old baby die with myelomeningocele. There are two cases of spina bifida in the immediate family history and the counselor gave her a 4% recurrence risk. She had taken an anticonvulsant (phenobarbital). The counselor felt she wanted to be told she would have a normal baby, but he could not give her that assurance. Her husband does not want another pregnancy and they argue a great deal about reproductive choices. They had separated and reconciled, and were using contraceptives sporadically. She is afraid of getting pregnant, wants to be pregnant and have a healthy baby, is trying to adopt a child, and is considering a tubal ligation, which indicates that her decision-making is very fluid. She has ruled out testing with selective abortion as an option. Her opposition to abortion is firm but based on common misperceptions (saline abortion "burns" the fetus, a D and E pulls the fetus out piece by identifiable piece, abortion is horribly painful for the fetus).

I have only illustrated here the cases of those patients who believe they have reached a decision about abortion before the test results. There is another large group of cases where the decision still cannot be made at this point; they simply are not sure how they will decide if faced with the situation. Even among those who have made a decision about abortion before being in the actual choice situation, changes of decision occur. Some who thought they would abort, find they really choose not to, and some who had decided not to abort, when faced with the actual choice, decided to. Projections are not always accurate forecasts of the future, as I have argued in another context.[67] Because of this, a precommitment to abort actually subtracts from the full range of choices and can only be ethically justified under very extreme conditions or when it entails a choice for self-destructive consequences that those of us in the affirmative stance toward life do not have to honor in any event. In addition, decisions made before we are individually confronted with an actual situation tend to reflect our unexamined acceptance of cultural conditioners. Our particular interests and needs do not realistically intrude on the internalized patterns of our environment at this point in the decision-making process, or they do so in an abstract, nonthreatening way. They are more intellectualized than experienced. It is really with the imparting of information from a test, or from risk figures if no definitive test is available, that we become totally involved in the decision-making process. It certainly makes pragmatic sense to examine possible scenarios and our attitudes toward the available options. We can sometimes identify options we had not considered and we can do some work to dilute the stress of a negative situation so that shock is lessened. Such participation has considerable value and is something that a rational review of possibilities can accomplish. Philosophical tools are useful in this preliminary stage of stress for their predictive value, for the initial ranking of interests, for the inferring of alternative choices and consequences of those choices, for example, and in fact most philosophical positions can adequately handle this point in the process. When the information is imparted, however, the individual is now wholly self-involved, and this stress can be understood only by considering both the rational and nonrational components of that individual's existence. A philosophical system that concentrates only on the rational nature of humankind is too narrow to give an adequate analysis of the actual stress situation. At this point, we are back to the nonrational justifications of intervention, to attitudes toward pain and suffering, and to self-affirming or self-destructive affects. The decision, when test results have cen-

---

[67]Donella H. Meadows, Dennis L. Meadows, Jorgen Randers, and William W. Behrens III, *The Limits to Growth*, New American Library, New York, 1972.

tered it squarely on the self, cannot be an exclusively rational decision.

Abortion Decided Affirmatively or Negatively with Test Results

Case Study 112. Trisomy 9 Mosaicism

This was a very much wanted pregnancy. The wife is in her late thirties and it is a second marriage for her. There are no children. The husband had had his vasectomy reversed, so a great deal of effort went into the pregnancy. The amniocentesis test indicated a trisomy 9 mosaicism. Only one trisomy 9 case has so far been prenatally detected, and this may be the first mosaicism. The abnormality results in early death, with 8 years being the outer survival limit thus far, and the effects of the mosaicism could include bone abnormalities, heart defects, and mental retardation. The lab was reasonably sure that this was a true mosaic, rather than a pseudomosaic, since it was apparent in two of three cultures.[68] An abortion was decided on, the process taking over 24 hours. The couple were together during the procedure and held the dead fetus when it was finally delivered. A nurse unfortunately intruded during this grieving process and could not understand the appropriateness of it. She was told to leave by the couple.

Case Study 113. Anencephaly

This was a maternal age counseling, with the woman ambivalent about the pregnancy. It was actually not much desired. Ultrasound examination indicated that there were twins, one normal, the other anencephalic. It was decided to attempt two taps, but separate sacs could not be located. The AFP level was very high for the affected fetus and abnormally high for the normal appearing one as well. Because the pregnancy was not strongly desired, the woman decided to abort. Autopsy indicated identical twins, one anencephalic, one normal.

Case Study 114. Duchenne's Muscular Dystrophy

There is a family history of Duchenne's disease, the wife's two brothers and her three uncles having died from it. She had a 50%

[68]New York State Chromosome Registry Meeting Minutes, May 9, 1979, Albany, New York.

chance of being a carrier, was tested, and her CPK was normal. However, the accuracy of this test is problematic. Amniocentesis was done and indicated a male fetus. She decided on an abortion. The couple have no children and this was the second selective abortion.

Case Study 115. Meckel's Syndrome

I previously discussed the first part of this case (#44). After deciding against maternal age testing, this couple delivered a baby with Meckel's syndrome who died at birth. They were counseled and became pregnant again. Ultrasound and AFP test results were ominous. The fetal head was not growing, amniotic fluid was red, general size was small, and AFP was very elevated. The couple had a conservative attitude toward abortion. However, the prognosis for the fetus was not good; it would probably not survive to term. A second ultrasound test showed a probable ruptured encephalocele and the parents decided on abortion. They are very discouraged.

Case Study 116. Fetal Abdominal Mass

Because of uncertainty about gestation dates, an ultrasound test was done. The fetus showed a large abdominal mass filling almost its entire abdomen. It was feared that this was inhibiting fetal growth and might interfere with delivery. The obstetricians considered and rejected attempting to tap it. Cultured cells from amniocentesis were growing poorly. The couple was highly motivated to continue the pregnancy since it was very much wanted. Everyone decided to wait. There was the possibility the fetus would die, and the possibility of risk to the mother, The pregnancy was carefully monitored. The baby was delivered by cesarian section and had numerous problems. The spine was deformed, there was a urine-filled sac (megaureter), extremities were discolored, and the baby had an imperforate anus. However, the child went home, with corrective surgery scheduled.

Case Study 117. Down's Syndrome

One of the first Down's cases prenatally detected was carried to term after the parents elected not to abort. They found the information very useful, however, in helping them prepare themselves for taking care of their baby. They regularly attend a parents' group for the Down's problem, and now recommend to all parents that they use the resources of the prenatal program.

Many of the decisions for aborting were made by couples highly motivated to have children, and represented very tragic choices. Although a less strong commitment to the pregnancy process might render such a choice less stressful, there are practical reasons why abortion can never be a casual choice. The lateness of the choice is an important one—pregnancy is not simply a time span between conception and birth, a gap temporally characterized. Judith Jarvis Thomson's classic paper on abortion makes that mistake in characterization in her examples.[69] It is a physiological process involving the woman's total system and a process that is purposeful in its goal. Changes are effected in the system; the effect is systemic. That, in fact, is one of the reasons why forcing a woman to be pregnant for nine months is more than a bit of serving time and is, in fact, a major assault on her self. Changes have been made to her by force that are major changes and that have psychological correlates, at the very least. It amazes me that this merely temporal characterization of pregnancy has not been strongly disputed by women since we surely know better.

When abortion is elected at 5 months, awareness of the pregnancy (quickening, e.g.) is already very high, so that it is indeed a major choice. The labor process is a significant trauma and reverses the well-developed physical processes preparing for mothering in a physical sense. The termination of those processes may require considerable followup support. Before I discuss this last step in the decision net, however, I would like to review the nonrational elements that enter into ethical decision-making at this present step.

How we justify intervention, actually, how we feel about it, is central at this point. If our attitude is to accept suffering rather than attempt to change it, terminating pregnancy is interventionist and not an option. If we define ourselves in terms of resignation to controlling forces, growth through bearing of pain, life in a sense as a test of systems tolerance (how much pressure can be endured before the systems break down), then active nonacceptance is not an option. As I indicated in Chapter 1, this attitude toward pain and suffering is not rationally adjudicable, all arguments from stoicism to the contrary. It is not apparent that acceptance of suffering is a more central human role than intervention in the process, or that defining suffering as an instrumental good (choosing it) is any more rational a choice than defining it as bad and

[69] Judith Jarvis Thomson, "A Defense of Abortion," *Philosophy and Public Affairs*, 1 (No. 1), 47 (1971).

therefore to be prevented (not choosing it). Whether empathy and active anxiety, or anesthesia and calm is the better attitude toward existence is not a matter to be rationally settled, because the choice finally depends on valuing adaptation through change over a long span of time, rather than adaptation through static maintenance over a much shorter span. This choice, in turn, depends on affirmative attitudes toward existence, and these are all nonrational affects.

Those who decide on aborting the pregnancy decide on assuming some control over external circumstances through attempts to change those circumstances. Those who decide to allow the pregnancy process to continue to completion decided on controlling their own responses and accepting external circumstances. How individuals view intervention, then, affects how they will ethically view killing, and more specifically, homicide. Once we have developed our stance about intervention, we can then make reasonable estimates about increasing pain or decreasing pleasure, the cost/benefit of killing. Humans are *at least* as interventionist as accepting, and killing is therefore theoretically as well-based as nonkilling. Humans are *at least* as characterized by the efforts they exert to control their lives. To the extent that we describe ethics in terms of human responsibility for choices, ethics embodies interventionist choices quite comfortably, and has meaning only in such terms.

The decisions in the the case studies have also been made in terms of the interests and needs of the pregnant "persons." These interests become focused on a decision about the pregnancy process and are made, not in terms of principles about killing other "persons" (there are no such social entities at risk here), but in terms of the individual's needs concerning the pregnancy and in terms of projects about future pain and suffering. Decisions can be made in terms of rules of thumb about homicide (whether it is unnecessary, capricious, will be self-destructive to the actor, will create a social climate of fear and insecurity, increases pain and suffering), and again, these are not decisions made in terms of an inherent evil of homicide, but in terms of evaluating the consequences of such an act.

The killing of "persons" has a special interest for us all as human beings because we are all conscious of its implications for us. It is a threat to any of us, as Hobbes rightly saw,[70] and importantly

[70] Thomas Hobbes, *Leviathan*, Penguin Books, Baltimore, 1956; and especially Leo Strauss, *The Political Philosophy of Hobbes: Its Basis and Its Genesis*, University of Chicago Press, Chicago, 1952.

because of our consciousness that it can occur, our understanding that that awareness occurs in other persons like us, and our projects about our future security. It is shared human interest not to be killed, for those in the affirmative stance, a prerequisite for all other interests and needs, and hence an overriding value in a great many choice situations. But I hope to have demonstrated that the fetus, as part of the pregnancy process, is not a "person" and an interest in not killing the fetus has no such overriding value in terms of the interests of *actual* persons.

Nevertheless, the decision for abortion (homicide in my terminology) is not a trivial one. This is particularly true because of possible consequences to the persons involved in that decision and not because of any ungrounded principle about the sanctity of all life, or reverence for all life forms. Before concluding my analysis of the ethics of abortion, I need to look at those consequences to determine whether the intervention really is justified.

Followup Support
for Abortion Decisions

Case Study 118. Down's Syndrome

The couple had decided on aborting a Down's fetus after amniocentesis had indicated it was affected. After the abortion, counseling was provided for them by psychiatrically trained staff. Grief and loss have to be worked out, since the pregnancy process involves projects that are also terminated (given up) with the physical abortion. This couple wanted to do anything they could to help others with the same problem. They would be available for parents' groups or for self-help groups.

I also discussed a case in this chapter where holding the dead fetus was a way of working through the sadness about having to end the pregnancy. This is available to our patients as well, because it is another means of supportive followup.

What are the costs for the couple, if they make a decision for abortion, that must be outweighed by the benefits to them? In this discussion, I will be emphasizing the costs of aborting, but I am assuming the reader will keep in mind the costs of continuing the pregnancy (the slow death of a trisomy 18 baby, the physical and mental limitations for a Down's child, the self-mutilation of a Cri-du-Chat child). For abortion choices in general, Julio Aray has given the strongest statement in his psychoanalytic review of the psychological costs to the decision-maker, even when the decision on ethical grounds was clearly justified.[71] An abbreviated list includes: mourning pathology in which suicidal tendencies, phobias, and subliminal alterations are exhibited; damage to the bodily and psychological ego; because of the intricacies of sexual self-image, vitiation of the creative process and capacity in general through sublimation; repression of the abortion experience, rather than its working out, a pathology favored by lack of visualization of the object after abortion; the occurrence of failure neurosis, in terms of the emphasis on procreation as a measure of self-worth.

This analysis is not unchallenged, of course. Natalie Shainess finds all of these putative psychological costs of abortion rationalizations to suit its legal status or theologically and historically conditioned views, rather than valid psychiatric analysis.[72] Her argument as it stands is philosophically invalid, since it amounts to an assertion, and perhaps in the case of the rationalization charge, a genetic fallacy. It happens to be true that psychiatric literature is rather conservative, following social trends rather than leading them (e.g., on homosexuality, divorce, and, yes, abortion). This still tells us nothing concerning the psychological effects of terminating a major physiological process. It is probably also true that much guilt response owes to theological and historical time-lag. This also tells us little about the responses to physiological change except that they are expressed in terms of internalized environments and could be expressed in alternative ways. What we do know is that a major physiological process like pregnancy predisposes the individual to exhibit certain behavioral responses, and if the pregnancy process is terminated, feedback mechanisms will take some time to alter that state of preparedness. In such a fluid, confused span, various inappropriate psychologcial responses can be expected and need to be taken seriously. When familial and social interactions are added to this situation, followup support is not unreasonable. Since homicide has occurred, some grief and working through of that reality is required. This does not imply that the homicide was wrong, only that it was a major life event and that this sort of event, even when a pleasurable one, has a significant impact that may be poorly handled by the individual.

[71]Julio Aray "Unconscious Factors in Induced Abortion," *The World Biennial of Psychiatry and Psychotherapy*, Vol. II, Silvano Arieti, ed., Basic Books, New York, 1973, pp. 412–428.
[72]Natalie Shainess, "Women's Liberation—and Liberated Women," *The World Biennial of Psychiatry and Psychotherapy*, Silvano Arieti, ed., pp. 86–111.

What concerns me, then, about capricious abortion, is not that the homicide is wrong (it is not), but that the individual choosing it is so poorly socialized and so unresponsive to the environment, that that person fails to view it as a significant choice. It says something about the actor: the narrowness of interest, the insensitivity, the triviality of purposes, the neurotic force of superficial needs. The real ethics of abortion involves our affect when considering it.

This case study (#118) illustrates the importance of visually and tactilely experiencing the situation. Similar psychological factors operate for a late selective abortion as operate for the death of a young baby, because of the physiological preparation for birth that pregnancy represents, but they are not the same because of the absence of social interaction in the former. Still, closure is probably most effectively carried out by not repressing the experience, by understanding on a primary sensory level what has happened. Followup support in terms of counseling may also be helpful, since this is a major event. The various psychological effects that concern us can be handled and do not begin to compare in difficulty to studies indicating, for example, the disastrous psychological and social costs of the birth and parenting of a spina bifida child[73]. What needs to be made clear is that an evaluation of the psychological costs of abortion is not an argument in favor of a universal prohibition against homicide, since it is irrelevant to that issue, but part of a calculus to determine whether homicide is the action of choice in a particular situation.

What I need to do now is take a more detailed look at premises 1–4, since I believe I established premise 5 in Chapter 4, to complete the argument (see my argument sketch at the beginning of this chapter). First, I do not find it too problematic to maintain that killing is not always felt to be wrong, is not always a bad action, because it does not always cause pain and can have good consequences. We kill plants to survive (to avoid the vegetarian objection that killing animals to do that is wrong, but that we can nourish ourselves without killing), and I know of no serious (or frivolous) plant rights movement. We kill microbes, we kill insects that threaten our food supply, or health, or territorial space. We kill rodents. If we did not kill embryonic animals or otherwise obtain animal as well as plant protein, we would not survive. We kill (indirectly) whole species by competing for habitats. Since our in-

---

[73]Edward Guiney, "A Question of Priorities," *Journal of the Irish Medical Association* **66** (No. 15), 401 (1973); J. Lorber, "Results of Treatment of Myelomeningocele," *Dev. Med. Child. Neurol.* **13**, 279 (1971).

terests and needs, our choices, form our value system, the terminating of other life systems is inescapably built into our ethics. Killing is an action that expresses our values. Since we exist in proximity to other systems, since interests and needs are different and not routinely compatible, since our interests and needs determine our value system, in conflict situations killing expresses our valuation. That is simply the universe as it is. Not only can killing be an instrumental good, then, but an intrinsic good, if we wish to keep this distinction. Eating (killing) is a pleasure. Unless we wish to confine ourselves to a diet of milk and blood, I know of no way of eating that does not involve killing, if not the adult form, the embryo form, plant or animal.

I think most of us would agree with this, but we still tend to want to separate ourselves from this condition of the universe. Our humanities tell us that *homo sapiens* is somehow apart from the rest of the biosphere (because of soul, reason, emotions, language—the distinctive feature has changed through time). We wish to define ourselves as a separate class, between the gods and the animals, so that no understanding of the rest of the biosphere can span that gulf and threaten us. But each distinction of our making has collapsed. The soul depended on theology and without that base in the supernatural is meaningless. There is no space for an analysis of theological arguments here, but the state of philosophy is that for every theological proof, there is a rational counter-proof, and thus no adequate philosophical analysis can be built on theological assumptions. Reason lost its throne when animal learning was demonstrated. We are not the only problem-solvers on the planet. Although our children surpass the other primates, their thinking categories develop over time and at the early stages are shared by other animals.[74] Emotions, expressions of social bonding, are common to many social species besides ourselves, as the ethologists have clarified and as pet lovers knew all along.[75] We have fallen back to language as our line of separation, but that battle is almost lost as well. The work with signing primates (Ameslan) indicates that either animals are perfectly capable of communicating linguistically in our style, or somewhat face-

[74]John L. Phillips, Jr., *The Origins of Intellect: Piaget's Theory*, Freeman San Francisco, 1975; Jean Piaget, *Insights and Illusions of Philosophy*; Guy Woodruff, David Premack, and Keith Kennel, "Conservation of Liquid and Solid Quantity by the Chimpanzee," *Science* **202**, 991 (1978).

[75]Vitus b. Droscher, *The Friendly Beast*, Harper & Row, New York, 1970.

tiously that our deaf and mute individuals are not human be-ings.[76] We all know that Ameslan is an acceptable language capa-ble of artistic communication as well as everyday functions. We are, then, special to each other, and our interests and needs are special to us, but we are not so separate and special to the rest of the universe.

Homicide, therefore, is a subset of killing and concerns us to a greater degree than general killing, but can also be the right thing to do, can have good consequences. Because of our personal inter-est in homicide, our empathy, and therefore the threat to our exis-tential selves it represents, homicide is an instrumental good, but probably only psychopathically gives anyone pleasure or is an in-trinsic good.[77] However, again like killing in general, our social existence implies homicide. If we wish to drive cars, cross bridges, support present population size at current food prices, in short, at-tempt any complex construction of an environment (what we call civilization), we will pay with human lives. I am not necessarily talking about high technological levels of civilization. Bartering ex-peditions assume risk to human lives, the use of fire assumes cost in human lives. These are indirect homicides (even murders), but as I pointed out, direction or indirection is concerned with as-signing responsibility, not with evaluating the consequences, ex-cept to the actors. A human being is just as dead indirectly decided as directly decided. Although the homicide was not chosen for reasons of direct personal relationships, to tell a coal miner dying of black lung disease that there was nothing personal in our col-lective choice to kill him for the energy from coal is to say a very peculiar thing that is not relevant to whether it was the best choice or not except in terms of its effect on us, not on him. But then, pe-culiar things happen with collective responsibility, since it tends to dilute responsibility to the vanishing point. It does not change the question of the choice, however, or whether good or bad conse-quences have resulted. While collective responsibility makes as-signing responsibility a very difficult thing, of course, the fact re-mains that homicide, and in the case of a person, murder, was done. The method used was not as simple as the "smoking gun," but it involved setting up an environment in which an individual

[76]Eugene Linden, *Apes, Men, and Language*, Dutton, New York, 1970.

[77]Nikola Schipkowensky, "Epidemiological Aspects of Homicide," *The World Biennial of Psychiatry and Psychotherapy*, Silvano Arieti, ed., pp. 192–215.

would die, with knowledge that this would happen (a trade-off), if not knowledge of which specific individual it would happen to. Homicide, like killing, can be indiscriminate.

Because of our special interest in ourselves, an ethical justification of homicide requires stronger interests and needs than does a justification of killing. Ethically justifying murdering a "person" requires even stronger ones because the threat to our own existence as persons is more serious, more threatening to us. Even here it can be done, but it is not relevant to our discussion of abortion since this is now on the social level or organization, of "persons."

Let me finally restate the argument:

(1) General "killing" is part of the definition of, or necessary to, our existence; thus if we have an affirmative attitude toward existence, killing is an intrinsic and/or instrumental good.

(2) Since human beings are part of the biosphere, "homicide" is a subset of general killing, differing only in terms of human interests.

(3) "Homicide" is necessary to our social existence and is evaluated in terms of our interests and needs. Because of our special interest in ourselves, one of the consequences is that "homicide" (and more strongly "murder") can be threatening, and stronger needs are required to ethically justify it; but this restriction remains based on our interests, not on a principle prohibiting homicide, and is often satisfied.

(4) Abortion is "homicide" (and therefore "killing"), but is not "murder". "Murder" is a social level term describing the killing of a "person," a term that itself is on the social level and implies the expression of social interactions of at least the most rudimentary sort. "Homicides" are ethically justified on the basis of those strong interests and needs that can override our empathy with members of our species and the threat to our species' security this act represents. "Murder" is occasionally ethically justified, but only on the basis of much stronger needs and interests (what Schipkowensky calls "ultimate necessity") that must override our empathy with other "persons" and the threat to our own persons it represents.

(5) Therefore, persons can make decisions to abort on the basis of their interests and needs, and their projects, and in many situations abortion will be at least an instrumental good—and in terms of a broader context, part of an intrinsic good. It is not the lesser of two evils, but a good.

Summary

This chapter has not been a pleasant one to write. Killing does represent a threat, but it is of course a threat to our childhood wishes of omnipotence, security, and immortality. We really do not wish to acknowledge the unreality of that view of existence and our ethical systems have reflected that. The human condition, however, is not so wish-fulfilling as we would like, and the termination of complex systems is an integral part of that system. On occasion, since as human beings we are frequently interventionists, we assume some control by choosing termination, and then attempt to repress what we have really chosen and rationalize that killing another human being is always an evil to be avoided. We tortuously maintain this even in the carnage of war by concentrating on indirection, motivation, and responsibility, thus ignoring the consequences of trenches of human bodies or vaporized shadows. Ethical systems based on such a repression mechanism can scarely be useful in the actual, concrete situations, unfortunately. With the problem of abortion, the repression has had high human costs, costs to persons whose personal projects have been irremedially and unnecessarily damaged.

Hierarchy theory can help us avoid the conceptual traps of the abortion problem by theoretically grounding the concept of "person," removing its arbitrary and stipulative status. Again, the interventionist attitude is a central factor in the abortion question, underlining the affective elements in a proper ethical analysis. In addition, a description of the affective state of the decisionmaker in an abortion option is an integral part of the ethical analysis, the real ethics, actually, of abortion. Finally, actual "persons" make decisions to abort on the basis of their interests, needs, and projects, not in a simplified external calculus of consequence, but in a much more complex understanding of internal and external consequences and social nature.

Last, "homicides" are ethically justified on the basis of strong interests and needs that override our species empathy and threat to species' security.

Chapter 7

Social and Individual Interest Conflicts

One of the continuing, unresolved problems in medical ethics is the conflict between micro-ethics and macro-ethics, or individual and social interest conflicts. The individual, as a subsystem of the general social system, interacts with all other individuals in a precarious balance (controlled state or homeostasis) whose result is the social system. Not only do various individual needs and interests conflict with each other, but the specific needs of an individual (subsystem) can seriously conflict with maintaining the balance of the global system. One of the major problems in ethics (and politics) is the question of how we choose between these very different needs and who does the choosing. We know, for example, that complex feedback mechanisms in biological systems routinely function to sacrifice subsystems in order to maintain homeostasis. Shock in a complex organism is a good example of certain built-in physiological valuations of what is expendable in a trauma situation.

However, the characteristic of homeostatic systems to sacrifice subsystems in order to maintain stasis is not the only built-in conflict in our social system. If we switch perspective from the form-maintaining or structure-maintaining purposes of the global system and its effect on the component subsystems to the purposes of the individual subsystems, we gain quite another view of the conflict. What will satisfy the purposes of the subsystems, benefitting their needs and interests when looked at from the individual perspective, will often be at best neutral and frequently destructive to the purposes and stability of the large system. We can, for example, bring modern medical benefits to the Tarahumara Indians living at subsistence level on barren Mexican highlands. The high mortality rate for their children can be considerably lowered. Of course, the land would no longer support the additional population, and the impact of a high technology culture would result in the destruction of a thousands-year-old balance that their social system has maintained. "They" (in the large system sense) will be destroyed while "they" (in the individual sense) are being saved.[78] Drilling wells in the Sahel desert increased basic economic benefits to individual cattle herders, but left the system unable to support the increased herds under easily predictable drought conditions, eventually causing economic collapse and local famine. Garrett Hardin's "Tragedy of the Commons" is another example, where the pressure of individual need for economic growth collides with a finite "common or community land," damaging the shared resource.

The problem might be capable of solution if individual subsystems and the general system were separable, but of course they are ontologically the same thing. Only the perspective, the level of organization, the complexity, or the nature of our interest changes. In order to maintain balance or stasis, systems incorporate negative feedback mechanisms. In theoretical terms, such mechanisms do not sound threatening, but they translate into sickness, starvation, repression, eminent domain, environmental contaminants, and harsh decreases in standards of living among other niceties. The components these mechanisms impact on, however, *are* the system, and serious impact, with resultant individual destruction, can result in such wide oscillations that the system itself will disintegrate from internal stress. The danger points to the system's ability to maintain homeostasis occur at the oscillation peaks where stress may over- or undercorrect to such a degree that the pattern cannot be reinstituted. So the general system is always at risk to the effects of serious disruptions of the subsystems, even when these disruptions are caused by mechanisms whose purpose it is to maintain the general system. There is always the possibility of internal collapse. People die from shock, even though it is the physiological system's best try at maintaining its functioning structure.

At the same time, individuals are components of the general system and its fortunes are theirs. We *are* a social species and some social system is part of our definition of ourselves. We cannot avoid being affected by the disruption or destruction of our social system, which is why revolutionary change succeeds only where a system is already breaking down and why its costs are so high. The cost of evolutionary change is not minimal, either; the more modest payments are simply required over a longer period of

[78]L. Jolyon West, *Op. Cit.*

time. If the general system, through a consideration of individual purposes and interests only, does not adapt to the real environment, there is always the risk of collapse from external forces. In the Sahel, as the economic/ecological system collapsed, individual people starved to death.

So the macro-ethics (social)/micro-ethics (individual) problem really *cannot* be solved and we should not delude ourselves with philosophic optimism. A system must avoid both internal and external collapse, and the truth is that these purposes cannot always be compatible. Life is a very precarious balancing act. Let me suppose the most extreme examples of success at either internal or external stability, forgetting the costs of the negative feedback mechanisms required to achieve such successes. If a truly monolithic social system could exist (and I have my doubts about that possibility, but let me grant it), one that through feedback controls had imposed a truly static pattern of response to internal and external stimuli, one of the results (in fact, a desired result) would be that no new pattern of response could arise. Certainly internal oscillations would be minimal. This internal stability, however, is bought at the cost of inability to respond to new situations in the external environment. Evolution demonstrates that systems need to adapt responses to changing environments or they will collapse under external pressure. We know that the universe changes, and so there can be no permanent balancing act. A system that cannot generate new responses to new external pressures will fail. However, an internally stable system cannot generate such new responses by definition, and cannot achieve permanence or its goal in this way. The methods by which a system prevents internal collapse are not compatible with those required to prevent external collapse. The static society of the Tasaday, unable to self-generate new responses to a changing world, would be overrun by the Philippine logging industry, if not protected by outside societies (a sort of human game park). Utopian communities also tend to be short-lived.

The reverse is also true. Let us imagine we are attempting to construct a system of constantly generated new responses to the environment. In this case, because of the absence of some predictable patterns of response, the internal workings of the system would be chaotic. There would be no social structure in a pure example, but even in a system with the bare minimum of patterns of response, while adaptation to change could easily occur, internal instability would collapse the structure. The methods a system uses to prevent external collapse are not compatible with those required to prevent internal collapse. Any committee member knows the internal instability and lack of functioning of a free-form committee.

I have argued that ethics is a question of human choices and purposes. For social species, those choices and purposes involve the two perspectives I have been discussing here. Therefore, an ethics that would not involve conflict and trade-off or tension between fundamentally opposed purposes is impossible. What we are left with is the determination of necessary and unnecessary sacrifice, and an interest in maintaining a working balance, a reasonably equal tension, between social and individual purposes. This can be justified if survival is a value for us, because if either purpose is emphasized, the system is in danger of collapsing, either internally or externally. Conflict is built into my ethical system because it is built into the natural operation of systems.[79]

Our prenatal detection program exhibits the same interest conflicts in a number of specific areas. Again, some case studies will be helpful.

Sex Determination and Selection

Case Study 119. Request for Sex Selection

The center counseled a couple in their early thirties (younger than the 34–35-year-old rule of thumb cutoff for maternal age testing) who had three sons. This fourth pregnancy resulted from contraceptive failure and was a "totally unwanted" pregnancy. The parents requested amniocentesis specifically to determine sex. Although they stated that they abhored abortion, they would terminate the pregnancy if the fetus were male, or if the fetus were abnormal. They also did have concerns about possible chromosomal abnormalities, although with negative family history and both in their early thirties, they have only the general population's average risk. The main concern is that the pregnancy was unwanted,

---

[79]This is an answer to Robert Veatch's more optimistic view of conflict resolution, which is erroneous because it does not consider systems dynamics and hence feels that only individual conflicts occur, or that justice is an achievable ideal within a functioning system. See Robert Veatch, *Case Studies in Medical Ethics*, Harvard University Press, Cambridge, Mass., 1977, pp. 83–88.

but that they would choose to continue the pregnancy if the fetus were female. They also were firm that if no test were done and no information given them, they would terminate the pregnancy. A psychiatric consultation was acceptable to them and the psychiatrist confirmed their statements: they would abort if no test were done; they were very upset about the pregnancy and strongly desired not to have another son, and they could accept the pregnancy if the fetus were female. The wife was under great stress and was very firm about the decision to abort without the test. It was recommended that the parents be allowed to make their decision with the test information, to have the opportunity for an informed choice, and that the parents' psychological state was consistent with performing the test. A great deal of time was spent with this couple. Their expression of their needs and interests had been honest and not manipulative, but in some ways this created more initial problems for the counselor and the program than if they had manufactured a positive family history or simply expressed anxiety as their main reason for requesting the test. Even if doing the test were the right decision for their individual needs, it would place the welfare of the program itself in some jeopardy in terms of the social context in which it operates.

Case Study 120. Maternal Age

An Oriental couple, at the wife's age of 34 on the border for age testing, were counseled for an unplanned pregnancy. They have three daughters. Although not happy with the pregnancy, they said they would not want to abort unless something were wrong and did not seem to focus on concerns about the sex of the baby. The counselor, however, is wary about sex bias selection with Oriental patients. The cultural valuation of male children leads the counselor to ask more questions about desire for a child of a particular sex than he might for other couples. The purpose is not to identify sex selection as a motive *per se*, but to attempt to elicit the real reasons and motivations of the couple, primarily to thereafter be better able to help them. Again, however, if sex selection is revealed to be the actual reason for requesting the test, significant problems will arise.

Case Study 121. Maternal Age

Since the wife is 37, there are age indications for amniocentesis over and above any concern with sex preferences. The couple has two daughters and comes from a social milieu that highly values males. In fact, they very much want a son, not another daughter. Because of a smallpox vaccination while pregnant before this current pregnancy, they have already experienced one selective abortion. In this case, sex selection could be an additional result of the maternal age test, but the test would be done on the basis of acceptable medical indications for it.

What are the problems with sex selection? It is not illegal, of course. An abortion can be chosen by a competent woman up to 6 months gestation for reasons the woman feels are valid for her without violating any law. Although a legal right, however, there is strong social disapproval for such a choice based on the woman's interest in having a child of a particular sex (it is, by the way, not always a male child that is desired). A survey done in our center's regional area produced maximum anti-abortion responses to the question: Should a woman have the right to an abortion because the fetus is not the sex she desires? Only about 10% of responses in the first study would allow this.[80] If the program were to routinely do amniocentesis for this motivation, it would operate in the face of strong social disapproval and negative feedback pressures, many of which could jeopardize the effectiveness, and perhaps even the existence, of the program. Regardless of the individual merits of the case, the individual's interests are in serious conflict with the larger social interests of the program (to function effectively socially). Should individuals be sacrificed to these social interests? The answer, given the realities of such built-in conflicts, must be: Yes, if necessary; no, if unnecesary. The problem, as I previously stated, is to make accurate decisions about what is necessary.

There are some interesting side issues to this. One is that two approaches can be used regarding the conflict: (1) to attempt to lessen as much as possible the impact of the conflict on individuals, e.g., by allowing psychiatric excuses for acting contrary to social requirements and pressures; by compensating in another manner for denying certain interests, such as when we give large cash settlements for exercise of eminent domain; by offering special support services such as retraining grants for those suffering as a result of social decisions—in short, by employing those buffering mechanisms a society develops to ease the social/individual conflict and confrontation, and to indirectly maintain social stabil-

[80]Richard Doherty and Klaus Roghmann, "Attitudes and Acceptance of Prenatal Diagnosis among Women and Physicians in the Rochester Region," *Medical Genetics* 323 (1979).

ity; (2) to reject palliative measures and force a confrontation in order to disrupt the social system sufficiently to obtain major or radical change; in other words, to require enough individuals to suffer enough pain that the previous social balance will have to be changed. The two approaches, in fact, illustrate again the two perspectives, the former the individual perspective and the associated attempts to achieve good consequences within that framework; the latter the social system perspective and the associated attempts to modify its goals in order for good to result. The latter characterizes, in fact, the Marxist approach, and within that framework the former response is frequently called a band-aid approach. The problem is that such a social perspective alone destroys the necessary tension between the two and sacrifices without much compassion or concern numberless individuals for the broader purposes of the social goals.[81] It assumes that a social system can be established first, then maintain itself in rather radical flux without concern for the myriad disruptions on the individual level, and that such individuals have primarily the role of expendable, spurred actors. And that this process need be our only concern. Such a model may work to precipitate a system's disintegration, if that is the goal. Whether the disintegration would have been necessary is the question that is begged in setting up this model, and this is really the important issue. *Were* such insisted-upon individual sacrifices necessary? In addition, the model for a system's internal breakup will not work for the establishment and maintenance of a system, since no system can extravagantly sacrifice its components without internally collapsing. The latter expectation is precisely what seems counted on in the second approach. But once we try to reconstruct a system, we need a new model and *not* one that guarantees systems disintegration but systems buildup. For that we need the social stability that paying attention to and satisfying individual needs and interests generates. Let me be clear. It is sometimes necessary to sacrifice individual interests, but it must not be done lightly or by appeal to a general principle requiring the sacrifice of individuals, and it must be done only after careful determination that it was really the only alternative. Sacrifice of individuals will not work as a general principle because the tension between the individual and the social orders is required to keep a system in balance at any temporal point, and such

---

[81]For an example, see Seymour L. Halleck, *The Politics of Therapy*, Science House, New York, 1971.

a social homeostasis cannot be abandoned as the revolutionary fervor strikes us.

The second issue is the ethical justification of sex selection, regardless of the social mores rejecting it. Again, there are alternatives: (1) sex selection is never ethically justified; (2) sex selection is a wrong human choice, but autonomous parental choice is an overriding good; (3) sex selection in itself is ethically justified. As is apparent, the second alternative is a combination of the other two, a compromise position that I find incorrect on both counts. I will argue:

(1) That sex selection can be a right or wrong human choice depending on the situation, and needs to be judged on the basis of the needs and purposes of the individuals considering that choice, not on questions about healthy vs unhealthy fetuses.

(2) Autonomy of choice is not a value that overrides all others.

This middle view, which I cannot help viewing as a bastard child, is well-represented by Powledge and Fletcher's Guidelines for Prenatal Diagnosis Programs in the *New England Journal of Medicine*[82]:

> Although we strongly oppose any movement aimed at making diagnosis of sex and selective abortion a part of ordinary medical practice and family planning, we recommend that no legal restrictions be placed on ascertainment of fetal sex. We think such restrictions would be ineffective and impossible to administer, would lead to subterfuge and, more important, would violate our objective of non-interference with parental choice, even when we disagree with that choice. Prenatal diagnosis is not now widely available for this purpose; indeed, amniocentesis is often refused to women who request it for that reason, largely on grounds that the procedure is expensive and possibly risky, and should be reserved for grave medical conditions. Though we support the right of individual physicians to refuse to perform prenatal diagnosis for sex choice, we also recognize that in special situations, sex choice can appear to parents to be justifiable. We think most couples should not seek such information, however. Discouragement of this use of prenatal diagnosis, by pointing out that the risks and stresses of second-trimester abortions are not trivial, will mean that such cases will at least not be very great in number, though availability of earlier sex-ascertainment technics, now in development, is likely to expand them considerably.

[82]Tabitha M. Powledge and John Fletcher, "Guidelines for the Ethical, Social and Legal Issues in Prenatal Diagnosis: A Report from the Genetics Research Group of the Hastings Center, Institute of Society, Ethics and the Life Sciences," *New England Journal of Medicine* **300**, 168 (1979).

Obviously, parental choice does not override good medical practice, as I think has been demonstrated throughout this volume. If a second-trimester abortion is so risky that physicians must see benefits that would balance that risk, and if sex selection is not such a benefit, then it is the wrong choice as well as not an overriding exercise of autonomy, since it is self-destructive behavior and cannot be ethically justified. On the other hand, if benefits to the parents do outweigh risks, then not only can sex selection "appear to parents to be justifiable," it *is* justifiable. To justify the medical intervention is also to justify the choice. The emphasis in the guideline is on restricting the number of parents who would choose to make sex selection determinations, and the last sentence, suggesting that regardless of the circumstances there is something basically wrong with sex selection, gives the hidden premise away: It is not the risk, the seriousness of second-trimester abortion that outbalances the good consequences to the parents of sex selection, because even if new techniques can bring the choice to an early first-trimester stage, the tone of disapproval remains. The writers of the guidelines disapprove of sex selection, even with early abortion a possibility, and it has little to do with the risk of the procedure. Further, their support of parental choice is balanced by indirect coercion (cost, risk, physician refusal, and collective disapproval for seeking such information). We must look deeper than the compromise position.

Powledge has serious ethical reservations about sex selection.[83] She has difficulty emotionally accepting it as an option, for undefined reasons. She also feels it would have bad demographic consequences, and would foster sex role stereotypes and devaluation of the female. In our culture, I know of no empirical data to support either concern.[84] Parents tend to wish to select a child of the sex absent in their family group. Parents of five boys want a girl, of five girls want a boy. Or young couples express the desire for one of each, or for a brother to be a playmate of their son or vice versa. Some couples want a boy first, to be a caring older brother (significantly more couples do want the first-born to be male). But some want a girl first to be a caring, older sister. Fathers can express a desire for a son or a daughter. The variations are extensive and the blend will not give us some demographic nightmare. Furthermore, child-bearing phalanxes enter the mixture in different numbers at different years, so that patterns of similar choices (if

[83]Tabitha Powledge, personal communication.

[84]Charles F. Westoff and Ronald R. Rindfuss, "Sex Pre-selection in the U.S.: Some Implications," *Science* **184**, 633 (1974).

we had them) would not give us the feared homogeneity. I think the demographic concerns reflect our misperception about how monolithic our culture actually is, or how lock-step we are in our timing of reproduction. We have visions of second grade class-rooms, all boys. It is, empirically at least, doubtful. Our other misperception is the slightly paranoid view we women might have that sex selection involves choosing males over females. If the cul-ture is already highly sexist, it will reflect that value, but not create it. In fact, if choice operated according to that scenario, it would actually seriously alter the sexist values of that culture, since fe-males in one generation would become a scarce good, and scarcity in this case would enhance value considerably. In any event, eco-nomic issues are going to determine sexual equality, not prenatal selection of sex.

Finally, working through the artificial issues, is sex selection ethically justified? Recalling my argument for abortion in general, yes if the needs and interests of the persons involved are strong enough. The issue is to determine whether the short- and long-range interests and needs of the parents will be best met by such a choice. Is their project of a child of a particular sex such an impor-tant goal for them that serious damage will be done if it is not reali-zable? Since we must balance the physical and psychological risks of a 5-month abortion and the threat to personal security that hom-icide involves against the importance of this project, it must be a rather major goal or purpose in the couple's life. If it seems more self-destructive than self-preserving, the goal can be questioned and the choice will not be ethically justified, regardless of notions of patient autonomy. That determination is only partially made by the patients and is finally made in the human interaction of physi-cians and patients. Societies are actually societies of human be-ings, not of rules or laws, and it is in these human relations that the strength or importance of the individual's project or goal is deter-mined. Rules are generalizations made from interactions experi-enced in the past, but, as in the case of the problem of induction, they have no unambiguous validity for a presently occurring situa-tion. To concentrate on the rules, to try to determine experience "by the book," is to place more faith in our habits and culturally accumulated patterns of response than is warranted, and to se-verely limit the character of our human relationships.

It will not do to assume that sex preference can never be an important enough goal for parents, or that such projects are al-ways sexist or pathological. In some cases, this may be the case, but in others, it will not be. In either case, the determination is made on the basis of interests of "persons" on the social and not the taxonomic level. Sex selection is wrong if the parents' project is maladaptive, or destructive to their other projects and their continuing development. But it is not wrong because some rule or principle says sex selection is trivial or pathological in every case.

As a result, if we succeed in developing a technique for collecting and culturing fetal cells in the first-trimester, or detecting testosterone levels when a selective abortion is a low-risk procedure, the strength or importance of the parents' interests can also be relatively less. If we develop a preconception selective technique, we only need to monitor demographic effects to determine whether there are any unwanted consequences we had not anticipated. Extending human choice in this way may initially be slightly disruptive, although there is no data that this would be the case. In fact, if it became routine, neither sex might be overvalued.

In spite of the individual ethical justification for sex selection, however, it is not in the interest of our program to offer it. The social pressures that would be generated would threaten the effectiveness and existence of the program. A decision may need to be made to sacrifice the legitimate interests and needs of some individuals in order to continue to meet the interests and needs of a larger number of individuals. It is an unpleasant and common decision, one that characterizes the continuing tension in any system. One hopes that the individuals whose needs and projects have been denied will be buffered in some way from the undesirable consequences of such a denial. One also hopes that very important justified interests will in time accumulate enough pressure on the general system to effect change and adaptation, or at least reasonable responsiveness. This is a power-brokerage situation, but also an essential part of ethics. What may be the right choice from an individual's perspective is not always the right choice from a general system's perspective. What is always the generally adequate choice is that both perspectives must have sufficient power to keep the balance required for stasis. In this case, unfortunately, we know there is no rational social interest in restricting sex selection, but that there is an irrational emotive consensus to do so. A program that challenges such a consensus will jeopardize the good it can do. If the number of individual sacrifices significantly increases, however, the need to maintain a workable tension between social and individual goals will ethically require just such a challenge.

Excuse for Aborting
Unwanted Pregnancies

The sick role provides an excusing mechanism for those assuming it (with physician approval), a role that may sometimes be manipulative or an unproductive and inadequate defense mechanism. If the social system, or a cultural or religious subset of that system, disapproves of a choice such as elective abortion, individuals will attempt to adopt the sick role by entering a prenatal detection program. Should such a program serve this excusing function? Is it in the program's interest or the individual's interest to function in this way, and can social pressure mechanisms be handled in this manner?

Case Study 122. Mucopolysaccharidosis, Hunter's Syndrome

Both the 4-months-pregnant girl and her mother are confirmed carriers of Hunter's syndrome, an X-linked genetic defect causing severe retardation. She is unmarried, without support from the father, and this was an unplanned pregnancy that is not really desired. However, her family belongs to a religious group that opposes abortion, so an elective abortion was not an early option for her. Nevertheless, if amniocentesis indicated that the fetus were male, and if testing showed the fetus had Hunter's disease, she could abort in good conscience, since such a medical diagnosis would in her mind excuse her from the religious restriction and perhaps within her family system as well. Furthermore, this type of testing is precisely what such programs are set up to do, regardless of the motivations of the individual patients. Hunter's syndrome can be prenatally detected, is a severely retarding, debilitating, and progressively worsening disease, and medical indications justify intervention.

Case Study 123. Consanguinity

The dynamics of this case are very complicated. The 16-year-old girl is pregnant by a 21-year-old boy who ostensibly is not related to her at all, but actually is the equivalent of a first cousin. The boy's real father (he was raised in the family of his mother and her husband as her husband's child) is also the girl's grandfather. Neither the girl's nor boy's fathers are aware of the pregnancy and the boy wishes to leave decisions up to her. Her mother had made the appointment with the counselor and was very assertive during the interview. She was present for the full time with her daughter at the counseling session. The girl was not very verbal, and her mother was probably manipulating the session. The counselor later felt he should have seen both separately. The girl appeared to want to have the baby, if it would be normal. Her mother seemed to expect the counselor to give a very high risk figure for defects because of consanguinity, and hoped that would convince her daughter to terminate the pregnancy. The girl is still trying to decide whether to have an abortion. She and the boy will not marry. Actually, the risk figures for her are not as alarming as her mother had anticipated. From sibling to first cousin, the risk drops fourfold. The counselor informed her that the risk for the general population is 5%, and that their consanguinity risk is 10%. In addition, except for cases of arthritis, the family history is basically negative.

The girl's mother is probably attempting to use the program to force a positive abortion decision on her daughter. It would give an excuse for aborting, or more precisely, an excuse for the mother's coercion. One of the major questions in this case was what was the peripheral problem and what was the central problem: consanguinity or decision about abortion. If consanguinity were the central issue, the joint interview and the mother's manipulation of the program are not issues, since the genetics of this situation were carefully explained and the information imparted was not distorted appreciably by the mother's presence or manipulation. However, if the exploration of the girl's options and her perceptions about the pregnancy are central, then the program's interests and purposes have been affected by the mother's manipulation. Also, one needs to clearly define in this case who the patient is.

Much has been written about the physician's power to influence in the physician/patient relationship, so we sometimes forget the power of the patient to manipulate as well. The individual can use social levers, health care systems, and so on to advance his or her own interests, whether well or poorly conceived. Any health program can be manipulated, and basically, the major problem with this is that the manipulator is deprived of the honest, open interchange that could clear up misperceptions, identify additional options, or give real support in dealing with stress situations. Manipulation restricts the medical interaction to the patient's preconceived perception and deprives that patient of some of the physician's important experience. In some cases, an honest confrontation and pressuring could open up new alternatives and release the patient from unproductive fears and distortions. It certainly is more productive of human development to increase the range of responses. The manipulator wins the particular battle with social pressures, but loses the chance to grow into a person who can function more fully in relation to those social actualities. If it is wrong to use a medical program in this way, it is wrong because of the broad consequences to the user, not because of any institutional sanctity.

Regardless of the patient's hidden agenda, any medical program will justify intervention on the basis of reducing pain and suffering, and will expect to incorporate some denial and defense mechanisms in the interaction. Medical excuses are not always manipulative or definable as denial, of course. They are important buffering mechanisms for the necessary sacrifice of individual interests and keep the general system from internal collapse. Chaos and anarchy are no more important values than form and stability, so that such mechanisms are not intrinsically bad, and I do hope I have successfully argued that it would be a conceptual disaster to view them as such. Nor should we be eager to agree that the instrumental use of such buffering mechanisms so defuses the problem that change cannot occur or that their use is not itself a verbally disguised form of change. Casuistry sometimes can effect more change than direct confrontation and may be necessary to overcome the inertia of some complex systems. Relabeling can produce behavioral changes that over a period of time redefine the meanings of the labels. When enough excusing occurs, change also has occurred, even if it seems indirect. It is much more satisfying to cut the Gordian knot with a broadsword, but teasing it untied is more useful if we need the cord.

Referral and Feedback

As consulting, specialty services, prenatal detection programs depend on referral from regional physicians, and therefore require good relations with those physicians. But in addition, such programs deal directly with the individual patients, interacting with them and giving them some information directly rather than by channeling everything through the referral physician. This can create very delicate situations if information given the patient by the referral physician is inaccurate; if the patient prefers a continued interaction with the specialist and the center; if the specialist disagrees with the management of the case; if the referral physician has created in the patient some misunderstanding about what help a program can make available; if reports sent to the referring physician are late or too sketchy, or in general if communication is poor between referring physician and specialist; or if the patient is dissatisfied with the specialist's services and complains to the referring physician.

Many of these problem areas concern protocol among fellow physicians and are essentially concerns of professional relationship. However, some can generate conflicts between the interests of the patients and the interests of the program, specifically, in working well with the referring physicians. The nondirective, informing, patient-participating methods of most such programs as ours may also conflict with the patient-interaction style of the particular referring doctor.

Case Study 124. Possible Mosaicism

An obstetrician referred a 29-year-old woman for chromosome testing. This was her third pregnancy. She is quite short, although her entire family exhibits this trait. The referral was triggered by a suspicious buccal smear, done in the obstetrician's office lab. Such tests for sex chromosome abnormalities are unreliable and karyotyping is recommended instead. Her obstetrician assured her that everything was fine, and did not tell her that chromosome testing was the purpose of the referral and that the patient would be presented the option of chromosome analysis by the counselor. A report will be sent to her physician. In all cases, the program stresses the positive aspects of the referral since it deals directly with the patient. In this case, the referring physician, who is not a specialist in genetics, is not sure whether the buccal smear is a cause for concern and is thus seeking expert advice. There will be many cases where the specialist can easily determine that there is no cause for concern, but it will not be that apparent to the nonspecialist in the field. The non-informing method of the obstetrician varies from that of the counselor, but the counselor feels that first responsibility is owed to the patient and will make the decision on informing, and on the level of informing, based on that responsibility.

Case Study 125. Spinal Muscular Atrophy

The couple have a daughter who, some time after a febrile episode, lost her skill at standing and now exhibits progressive atrophy. A muscle biopsy indicated spinal muscular atrophy, probably the Kugelbeg-Welander type (Type III). The parents are not clear on the prognosis for their daughter and the referring physician has been nonspecific about that. The counselor knows that the prognosis is not as good for early-onset as late-onset disease (although early adult survival is the best prognosis for this group of atrophies). The referring physician has not given them a detailed, step-by-step account of even the best of the possible prognoses. The counselor is not completely sure which type this spinal atrophy is, and even the type classification is theoretically not well-established. Should the counselor pursue the prognosis further than the referring physician has? The couple would like to have another child, but the recurrence risk is 25%. Is it necessary to inform them of the range of prognoses in order for them to make an informed decision? Would this negatively interfere with the referring physician's management of this case?

Case Study 126. Misinformation

This is a review of previously discussed cases, concentrating on misinformation given by the referring physician or garbled information that the patient believes was given by the referring physician. The husband who had porphyria had been told by the family physician that there was a one in a million chance of his children having it (the counselor was obliged to tell him it was really one chance in two). The couple who had had a baby with an encephalocele had been told by the physicians at the hospital that the recurrence risk was 1 in 2000 (the counselor had to tell them that the chance was 1 in 20). The woman with a family history of Duchenne's muscular dystrophy had been told by physicians that amniocentesis could prenatally detect Duchenne's, and that her chances of being a carrier were astronomically low. No reference had been made to the risk of fetoscopy (the counselor had to tell her that amniocentesis could not detect Duchenne's, and that a test had revealed she had a high probability of being a carrier; the counselor also had to make sure that she understood the much higher risk of fetoscopy with its 90% best-possible-accuracy rate). The counselor must make sure to present such information in a way that will allow the patient to develop trust; that will meet the patient's needs and interests, and yet will convey the fact that medical knowledge is not completely certain, and that physicians can make mistakes especially in fields beyond their specialty; that information is constantly accumulating and being modified; and that will not unnecessarily threaten the patient's working relationship with the referring physician.

Case Study 127. Perforated Uterus

As the physician referral network grows (a result in part of physician satisfaction and confidence in a program's services), our program experiences more hot-line situations (immediate and emergency referrals). The largest number of referrals eventually come from obstetricians confronting known problems, rather than from internists and pediatricians who identify potential problems. In this case, a 21-year-old woman, 9 weeks pregnant, called with a serious problem. She had decided on an abortion, but the obstetrician who attempted it in his office perforated the uterus and compounded this by attempting a curettage. The abortion attempt was terminated at this point, the physician's assistant was called in, the woman's condition was stabilized, and she was allowed to leave the office. Fortunately, no infection developed, but the counselor certainly felt she should have been hospitalized rather than being allowed to leave the office. She then contacted Planned Parenthood, received the names of other obstetricians, and found one willing to attempt an office abortion. She was quite frightened by this time, and the first physician's younger partner was recontacted. This doctor recommended that the procedure be done only in a hospital and offered to do it without charge. A laparoscopy that had been done had shown no damage, but hospitalization would be indicated. She was also referred to our program. The counselor talked with both the woman and her physician; each felt more secure about doing the procedure at the Medical Center. The specialist there would do an ultrasound study to make sure there is no uterine abnormality (which might explain the perforation), and will first try the prostaglandin suppository method for aborting, before doing a suction. The counselor is willing to be very directive in this case, in the role of expert, and the woman was not only advised against an office procedure, but the counselor would have called any obstetrician willing to do an office abortion and strongly advised against it. There is also concern here about the followup after the first procedure. The counselor would have hospitalized the patient rather than letting her leave the office. There is a time-lag between her having left the office of her original physician and her recontacting the younger obstetrician and the program. During this time the woman was uninformed, at some risk, and could have made a destructive choice (e.g., getting an office abortion). Now that the program counselor has entered the case, at the suggestion of the younger obstetrician, the patient's medical interests are the program's primary obligation, and if necessary the counselor would risk its referral relationship to meet that obligation. However, such a confrontation might be contrived and unnecessary. The program is working with the younger obstetrician as a resource, and the referral has in fact prevented the woman from mistakenly seeking an office abortion. The program might now decide to be directive by calling Planned Parenthood to better acquaint them with proper medical management in such cases, or perhaps simply to inform them of the situation. Should this be done?

The prenatal detection program needs to be seen as a positive benefit for the patient, ideally by all three parties involved: the counselor, the patient, and the referring physician, who usually will be giving the patient long-term, continuing care. Although this balance may not be possible, it is a continuing goal. When interests conflict, however, the expert role does require that the counselor make decisions in terms of a professional view of the patient's interests, leavened by a somewhat humble view of the intensity and duration of the counselor/patient interaction. The manner in which these decisions are implemented can usually be adjusted to avoid a direct conflict with the referring physician and any attendant dissatisfaction. The probability of this occurring with any frequency is low, and the effectiveness and existence of our program is therefore not critically at risk over this issue. Professional conflict does not present as severe a challenge as the sex selection problem might, and because it is an internal problem, there are more avenues for settling conflicts by attempting to reach a consensus on good medical practice (or agreed-upon, institutionalized values) in the medical community. The conflict of such a program with social mores is much less a matter of consensus or institutionalization of values and therefore not only appears to be, but is, a more direct value conflict. Both professional conflict and conflict with social mores, however, are value conflicts, though in the former there are agreed-upon overriding values that can resolve the conflict and therefore there is less need for pressure and interest confrontation.

The Physician in the Dual Role of Filling Social Interests and Individual Interests

The physician is becoming more frequently pressed into meeting social interest expectations of cost containment, beyond the traditional expectation of meeting the individual patient's interests.

Such a dual role can involve a serious conflict and is sometimes referred to in the medical ethics literature as the "double agent" position.[85] In a sense, this is an unfortunate play on words that will not help clarify an important problem that needs careful consideration. The term "double agent" contains a sub rosa (or not so sub rosa) reference to intelligence agencies, where the double agent is a spy whose loyalty actually belongs to the side supposedly being spied on, or is a spy with no loyalty to either side, but only to the highest bidder. The whole espionage metaphor, in fact, suggests a lack of trust. The physician, whether psychiatrist-as-double-agent or genetic counselor-as-double-agent, confronts an actual ethical dilemma involving the conflict of macro- and micro-ethics, and not one involving disguised loyalty or dubious personal trustworthiness, as the espionage metaphor subtlely suggests. It was an unfortunate choice of terms, and perhaps not so innocent at that. I prefer not to use it, but to keep the discussion centered in its systems perspective and on what role the physician should play in social/individual conflicts when these conflicts are unavoidable.

The most dangerous of these conflicts likely to affect the physician/patient relationship is that between the cost-containment social goal and the individual patient's medical need. If a physician is trapped in this conflict of interests, there is no satisfactory solution, an unhappy result that I hope will become clear after looking at some case studies.

Case Study 128. Menke's Syndrome

This is a rare, X-linked recessive disease. A family was identified by our program as having carriers the study of whom would have significant value in research on this illness. The basic metabolic problem is a lack of copper uptake by the system, resulting in very poor body temperature control, brain dysfunction, and eventual death. The gene is lethal by two years. Another feature is kinky hair, with a distinctive hair structure, perhaps also present in heterozygotes. Attempts at copper supplementation have been unsuccessful because the uptake mechanism does not function.

[85]Daniel Callahan, "The Psychiatrist as Double Agent, "*Hastings Center Report* 4(No. 1), 12 (1974); *In the Service of the State: The Psychiatrist as Double Agent,* Hastings Center Report Special Supplement, April, 1978; Fritz Redlich and Richard Mollica, "Overview: Ethical Issues in Contemporary Psychiatry," *American Journal of Psychiatry,* **133**(2), 125 (1976). Burr Eichelman and J. D. Barchas, "Ethical Aspects of Psychiatry," *Op. Cit.*; Seymour L. Halleck, *Op. Cit.*

This family had a previous baby die from Menke's syndrome, so the mother is an obligate carrier. They have another baby boy who also has the disease. The program would like to work toward possible development of a prenatal test. With that end in mind, the family was asked to come in and participate in research that seemed promising for such a test: for example, hair analysis, fibroblast studies, CAT scan, EEG, and urine and blood tests. If the program is concentrated on containing costs or on distributing benefits to the largest number of people for equivalent costs, then we must question the social value of research and test development on this rare disorder. If, however, we adopt the individual care perspective, then such research and development may prove very beneficial to those individuals carrying the Menke gene and help reduce that particular suffering in the world. Should the program concentrate on the most common genetic diseases and channel money only into those areas? Should society in general, through consideration of other than expert medical concerns, determine what resources should be made available for which medical problems? Should the individual physician contribute to cost-containment judgments concerning his or her patients?

Case Study 128A. Spontaneous Abortions

These are becoming more common counseling cases, in part owing to the developments in chromosome analysis (e.g., banding techniques), more frequent analysis of abortus products, accumulated data from amniocentesis, and more general public awareness of chromosome abnormalities as a cause of miscarriages. A 34-year-old woman was counseled who had had three spontaneous abortions and no live births. (I could have chosen many similar cases.) Her family history was negative. Chromosome analysis had not been done for any of the abortions, although they were all early enough that chromosome abnormality could be suspected as a cause. At this point, three early spontaneous abortions are medical indication for a chromosome analysis. Although the information provided by such a test can at least help an individual psychologically, should the program consider cost containment in deciding whether to do such an analysis, even if the patient requests it? Or should the individual benefits of arriving at a diagnosis be the program's chief concern?

Case Study 129. Infertility

This couple, from a culture placing great value on procreation, came in for counseling and a workup concerning infertility. For the wife, menses began at 19 and she has been irregular. They have been married and trying to conceive for only about a year. Preliminary workup involving sperm count, testing for ovulation, and so on have not yet been done. The cytogenetic studies proposed would be done on the wife. Should the program screen out such patients? In general, should a decision be made to curtail spending on infertility problems as a low social priority. (Tabitha Powledge, for example, has suggested that infertility research, including in vitro fertilization, be discouraged as a misusue of social resources, since overpopulation is such a critical problem.[86] This raises a number of problems, one being that contraceptive techniques arise from infertility research, and it is in fact almost impossible to foresee what research area will generate needed developments. The question of who will actually ration the resource is also vital in terms of its effect on the physician/patient relationship and on the development of medical science.)

Case Study 130. *Hydrocephaly, Reproductive Loss*

Counseling was requested to determine the risk of another unsuccessful pregnancy in a history of reproductive wastage. The first and third pregnancies had ended with early spontaneous abortions, the second was an ectopic pregnancy, and the fourth and most recent resulted in a stillborn hydrocephalic baby. It could not be determined whether the hydrocephaly was congenital. The couple is under very great stress. Their family histories are basically negative, but none of the wife's siblings is at reproductive age. The wife is very upset at the prospect of another reproductive failure and cried during the session. She is sufficiently frightened to be very reluctant to risk becoming pregnant again. The husband would like to try another time. The counselor tried to reassure them that their risk was not significantly greater than anyone else's. Amniocentesis was not recommended for them, but ultrasound tests to follow fetal head growth were suggested. Again, the normal risk of unsuccessful pregnancies was stressed and the counselor also tried to be supportive to the wife who was showing considerable psychological reaction to the series of traumas. A chromosome analysis was not suggested.

The counselor was making his recommendation based on balancing risks of intervention with benefits of intervention for the individual, and not on a cost containment policy for the program. This is a critical issue in the social/individual conflict. It has fre-

[86]Tabitha Powledge, personal communication.

quently been suggested that physicians be more cost conscious and consider social priorities in their practice. Although the medical expert may be in a better position to make a cost containment judgment and rank medical priorities, and although accepting such a role would give the profession greater control over its future, the physician/patient relationship does not benefit from the physician's assumption of this role. And as a result, the individual patient does not benefit. There is little mechanism except pressure in a social system for rationing services and ranking priorities. All systems respond to feedback demands and environmental adaptation demands. The individual patient, however, is in a poor position to influence any of those mechanisms, and even an appeal to compassion is dependent on visibility, access to communication channels, and esthetic considerations. If, in addition to playing a power game with such social mechanisms, the patient must play the same power game with the physician, maintenance of the balance of social/individual dynamics, which I feel is the only ethical solution to the conflict issue, is seriously undermined. If the individual physician implements social rationing and ranking, the individual patient will almost always become a sacrifice to that physician's understanding of social goals and interests or to the physician's acquiescence in the system's expression of those goals. I cannot see how individuals could possibly maintain their equilibrium in the balance of power generally found in such circumstances.

Furthermore, serious changes would occur in the physician/patient relationship that would have to affect the quality of medical care. Just now, an individual patient in our society enters into the medical interaction knowing that his or her interests and needs will constitute the physician's primary value and will be met in terms of expert interpretation of those interests and joint selection of options that meet professional standards of care. If cost or availability is a problem, the physician will become the patient's advocate and, as someone knowledgeable about the health care system, will pressure it in the patient's interest. That traditional medical relationship is severely threatened by interpreting the physician's role as requiring social rationing interests. It introduces an adversarial aspect into the interaction, and for the same reasons I am concerned that the "legalizing model" will interpret this interaction as adversarial, I am also very concerned that this social concern role will result in an adversarial physician/patient relationship. If the physician were truly a "body mechanic," as some would have it, an adversarial interaction would not be a critical difficulty, merely an unpleasant but minor one. But I hope I have demonstrated that the medical interaction, to benefit the patient, must be one whose major components are caring, trust, assurance in being vulnerable, adequate communication, understanding and empathy of interests, expansion of options, and advocacy for options that are professionally judged beneficial. Those purposes are not consistent in many instances with a social concern role, so that the assumption of that role can be justified only in extreme and carefully identified circumstances and must not be institutionalized so that it is assumed in routine and general circumstances. The cost of such a role assumption would be the destruction of a working relationship with the patient and would: (1) make supportive medical care in situations of terminal illness, psychological stress, or chronic debilitation ineffective, because these areas clearly rely heavily on the personal, caring aspects of the medical interaction; and (2) lower the quality of routine acute medical care, because of the large psychosocial factor in the etiology, diagnosis, and treatment/prognosis of such acute disease. Since medicine is practiced on individual patients and not on the social system, neither physician nor patient should allow a social concern role to be institutionalized in the medical profession.

There are some few disparate, extreme situations in which a social role is justifiably assumed by the physician, but they are justified on the basis of preventing clear and serious danger to other individuals in the patient's social network, and only when that danger is so great that it outbalances the denial of the patient's legitimate and nonself-destructive interests. All generalizations and rules of thumb are inadequate when applied to every present and potential situation, and this fundamental assumption about the proper invocation of the social role reflects that. Responses, to be adaptive, require flexibility and not preprogrammed, narrowly channeled patterns. Thus there is room in the ethical system I am proposing for choosing social interests over individual ones. What there cannot be room for, without insuring failure to maintain medical goals, is the institutionalizing of the social interest role. Many of the "double agent" case problems arise from this institutionalization, which then calls into question the physician/patient relationship.

Nevertheless, I stated that the conflict problem really could not be solved ideally, except in terms of maintaining the tension or balance, and I think this remains quite accurate. The physician, as expert, is more capable of ranking various medical interests to insure the effectiveness and progress of the medical enterprise than is a nonprofessional. Also, one of the requirements for professional effectiveness is reasonable control of the development of the profession.[87] In abdicating the social concern role, medicine also loses, as a subsystem within the larger system. And in general, any threat to the system's purposes and interests (and hence, its stasis) becomes a threat to the subsystems as well. This is a well-known argument, assuming some familiar forms:

(1) If we do not regulate ourselves more closely, social regulative institutions will.

(2) If we do not contain costs, the government will.

(3) If we do not ration availability of services, bureaucracies will be set up to do this, and will do it with greater harmful consequences to us and our patients.

The argument is also partially true. There is a real danger of more regulation, government-imposed cost ceilings, and agency control of the medical enterprise. What is forgotten in ethically evaluating the situation, however, is that the power-broker, balance of power, tension of any social system needs to be maintained if we are interested in maintaining the functionality of both internal and external systems. There is ethical justification for that tension in terms of such maintenance. Within reason, the proper role of the expert is to advance an area in ways that are beneficial to the general social system and to the individuals served by that area, but when a conflict occurs, the first allegiance of the expert (to maintain the balance) must be to individual interests as professionally understood, except in unique and extreme cases. That is the only way to insure the tension required for systems maintenance. Not to participate in balancing power *is also to threaten social interests and purposes,* only in much more subtle ways.

I have not solved Garrett Hardin's *Tragedy of the Commons,* only restated it in medical terms and with more emphasis on the possibilities of individual pain and suffering than he had considered. There is no solution to the *Tragedy* in organizational or political systems terms, because such attempted solutions do not reflect the real problem. The actual difficulty is insufficient resources—in biological terms, the impacting of population density on environment. There are natural and Malthusian solutions for this, but I have yet to see an organizational or social system solution. If you have only a 10-inch pie, and you need a 20-inch pie to meet individual needs, arguments about the number of slices, or the shape of slices, or who gets a slice of the 10-inch pie will not change the

[87]Eliot Friedson, *Profession of Medicine,* Dodd, Mead, New York, 1970; *Professional Dominance,* Atherton Press, New York, 1972.

ultimate reality of unsatisfied individual needs. There is no solution to that except (1) to genuinely reduce need (not artifically by redefining need) by reducing the number of individuals, or (2) to find another 10-inch pie. The temptation is usually to redefine need, but a genuine need, unmet, will continue to exert a destabilizing effect on the system. In health care, the cost containment tendency is to assume that many health needs are not genuine, rather than to address the real problem of resource depletion and population balance. To institutionalize this social planner's assumption is to betray the patient and accomplish little toward the solution of the problem.

If one cannot prevent a conflict, the only ethical recourse is to try to maintain an uneasy balance between social and individual purposes. If the problem exists, there really is no satisfactory solution in terms of an ethical system. We recognize that when we talk about dilemmas and tragic choices in medical ethics, but we have not honestly addressed it in our theories of ethics.

Summary

When the macro-/micro-ethics conflict is seen in terms of systems theory, its inevitability becomes apparent. It is a built-in, unsolvable conflict, except in terms of the maintenance of a working balance between the two levels of goods. Further, such a balancing imperative is hypothetical because it depends on survival (and well-functioning) having a value for us. There are real, tragic choices in ethics, then, that are not theoretically well dealt with in traditional systems. The rights model and the social concern role are both adversarial in affect and assume that a rational calculus can combine these two levels, and successfully adjudicate between them, without consideration of a working balance based on equivalent power.

Chapter 8

Pragmatics as
an Ethical System

Adaptation and the Pragmatic Criterion

As I stated in the very first chapter, there is a fashion in medical ethics—which is perhaps an inelegant way of saying that our intellectual responses are culturally preprogrammed or conditioned by the disciplines we have mastered. The disvalue of such fashions, of all preprogrammed responses, is that a once adaptive mechanism does not remain adaptive over long periods of time. Since adaptation is a concept I have used frequently, I would like now to take a look at just what is meant by it. In a sense, the pragmatic criterion (fruitfulness, workability, or survival value) is a conceptual restatement of the biological notion of adaptation. Since the second clause of my ethical hypothetical generalization depends heavily on such a pragmatic criterion, let me examine it from a systems viewpoint to discover what it implies for the comparison of various human perspectives on the universe.

Adaptation involves feedback from the environment (i.e., experience). It assumes a transaction between one part of experience (us) and the broader part that we know through ourselves as instruments (an epistemological subjectivism or constructivism), but that we obviously do not control, encompass, or create (a posited ontological realism). What the pragmatic criterion is based on is a pleasure/pain principle as a result of which we usually choose to avoid painful stress and to judge our activities in terms of their likelihood of producing positive feedback from the environment (success, cash value, well-functioning) rather than negative feedback (failure, penury, illness). Painful stress can of course run quite a gamut. There is the possibility of acute physical trauma: if I act as if levitation were possible and attempt to prove it from a highway overpass, I will quite literally bump up against the hard reality the old realists talked about. There is the only seemingly less dangerous possibility of conceptual confusion or wheel-spinning: Let me give a rather lengthy example of this because it nicely illustrates the trouble we get into when we fail to pay proper attention to empirical feedback. Christopher Lasch has recently joined the ranks of those categorizing our cultural period as exhibiting the psychopathology of narcissism or much more imprecisely, selfish, empty self-involvement, something Rollo May anticipated a number of years ago.[88] Although he employs the conceptual category of narcissism borrowed from psychiatric literature, he has not evaluated this conceptual tool in terms of any actual data. No epidemiological study was used to probe the environment and determine whether the feedback showed an actual rise in the number of psychiatric diagnoses of narcissism. Such feedback data could have been obtained from psychiatric registers. So we are using a category to attempt to describe experience without providing any means of determining whether that description will work, first of all. At first glance, this does not strike anyone as a life-or-death situation. Humans have been using such arbitrary stipulations commonly over their history. Why need our categories be adaptive? Let me develop the argument a little further. Lasch now takes this untested conceptual tool (which in psychiatry has a reasonably specific definition or meaning) and uses it as a model to describe and critique his culture. To do this, however, he must broaden the conceptual model, and this he does by equating the psychiatric categories of narcissism and borderline disorders. He can do so without arousing severe challenge only because of already existing problems with psychiatric conceptual categories, problems stemming from the use of theoretical tools that lack the ideal clinical feedback mechanisms. And finally, he suggests, without honestly committing himself to it, the equation of narcissism and schizophrenia. This muddled model is then applied to our culture for purposes of flagellation and *fin du siècle* predictions of deterioration or degeneration. The danger is that the culture may very well accept such a description as accurate and begin to act on it (one nation's president already has). Guilt, sin, self-sacrifice, abjection, already serious problems in the psychological development of many individuals, may become more intransigent. But all of this can occur without the palpable designation of

[88]Christopher Lasch, *The Culture of Narcissism*, Norton, New York, 1978; Rollo May, *Love and Will*, Dell, New York, 1974. See also, Colleen Clements, "Misusing Psychiatric models: The Culture of Narcissism," *The Psychoanalytic Review*, Winter, 1981.

any "sin" at all. The effect on people's lives of such an accepted cultural category may very well *not* be insignificant. There may be considerable maladaptation and pragmatic disvalue, but there will be no feedback loop from such experience to the conceptual category to indicate that the category should be changed or abandoned. The harm to individuals that can be done by conceptual miscategorization is one strong ethical reason why such categories need to be adaptive, need to be tested in and by experience. But the humanities, unlike the sciences, are not accustomed to meeting such requirements of responsibility. It is not part of their tradition, as it is with medical science, for example. There are no feedback mechanisms built into the humanities process to provide pragmatic accountability. And the danger is real.

We are not free, then, to choose any stipulation or construct any category we wish, if we value survival and a clear head. If we choose to call a whale a fish, there will be clear disadvantages, ranging from thoroughly messing up our taxonomy of species to being rudely surprised when a whale surfaces under our small boat to "blow" since fish do not need to "blow." The valuable (in terms of survival or well-functioning) aspect of science or empiricism is that some stipulations can be determined to be better than others based on this unavoidable feedback from an environment or experience of which we are unalterably part. Bracketing the natural world, as the phenomenologists did, is incompatible with general systems theory, and for those with an affirmative attitude toward experience, is unethical as well.

Medical science, then, involves tested and always additionally testable responses (adaptive behavior, the pragmatic criterion). It is, of course, value-laden. Cognition is value-laden; there is no descriptive, objective fact that does not reduce to a transaction between human experience and the environment humans operate within, and operate within in terms of their purposes and interests. But being value-laden does not imply arbitrary stipulation, or that all is only our wishing or choosing. We do not construct the world from nothing (*ex nihilo*) or ourselves from nothing (*sui generis*).

The Traditional Trio
in Medical Ethics

So medicine is empirical and the advantage of having ethics interact with it to form the new discipline of medical ethics is to refine our valuing and choosing, to determine whether our ethical generalizations or rules are really adaptive or are patterned responses that, through feedback, will show themselves to be maladaptive. Medical ethics should constitute an opportunity for the reality-testing of ethical concepts, and thereby become a significant opportunity for ethics. Thus far, it has been a missed opportunity. Traditional ethical theories have been imposed on medicine, imported whole cloth from the humanities, forming basically three repeating patterns, three preprogrammed responses. The current trio of basic concepts in medical ethics, as I have pointed out throughout this book, are:

(1) Natural law/natural rights/and derivatively, absolute social (human) rights.

(2) Person as end as well as object/respect for persons.

(3) Utilitarian consequentialism.

These traditional philosphical positions all in one way or another emphasize the rationalistic aspect of human experience to the exclusion of other important modes of experiencing. There are major conceptual problems with each of them and I hope to have shown that none of them speaks to the real issues in medicine, which center around the interventionist nature of medicine and therefore the ethical justification of intervention itself, and which finally depend on affects of affirmation or pleasure in existence, compassion or empathy, and non-acceptance of pain and suffering. Making this a principle of beneficence ("one ought . . .") as Frankena does,[89] misses the point by trying to convert the affect into, instead, a rationalistic principle, thus missing the emotive foundation.

(1) Natural Law/Natural Rights/ Absolute Social (Human) Rights

Natural Law is a view of existence as ordered, structured, combining in law-like ways or forms. If we search for an answer to the question "What is the foundation or ground for natural rights?" there are several alternative explanations or grounds that can be offered:

(1) Natural rights derive from a natural law, which is the expression of a reality we know or can come to know.

(2) Natural rights are intuitively known and nonproblematically agreed upon.

[89]William K. Frankena, *Ethics*, Prentice-Hall, Englewood Cliffs, New Jersey, 1973.

(3) Natural rights are implied simply by our humanity.

(4) Natural rights are derived from principles, or are principles, arrived at in a unique social contract situation.

(5) Given certain theological assumptions, natural rights are given by a supreme lawmaker.

Statements (1) and (5) both presume a metaphysics, without which such rights are foundationless. Since the concept of natural rights, derived from a supposedly reality-descriptive natural law, has the oldest tradition, I want to first look at the problems with this proposed ground for rights. In systems language, it stresses negative entropy and open systems, as opposed to positive entropy and closed systems. It is one of *two* views of the universe, seen both in the pre-Socratic philosophic tradition and the mytho-religious tradition. The second view is one of process, change, contingency, chaos, the breaking down of structure—all the characteristics, in fact, finally attributed to that recalcitrant, problematic stepchild of philosophy, matter. By emphasizing only the structuring and complexity of systems, while overlooking their equal if not greater tendency to breakdown and changing configurations, natural law gave us a static, unchanged ordering of existence as its primary quality, a serious distortion. We now know that negative entropy and open systems can exist and function because energy is imported into these systems from a positive entropy, closed system. Form or order is paid for by chaos. It too is contingent, in process, restructuring.[90] If we look for the "laws of the universe" or the "natural order of things," we will at best find them temporary, momentary, and cannot apply those "laws" as unchanging rules to the present experience. We fool ourselves into thinking we can step into the same river twice because for most of our purposes the changes are small enough not to much affect them.

We cannot derive natural law from existence as we know it, then. If a theological base is attempted, an above-existence lawmaker or order-imposer, we still have problems. Philosophy can not ground an ethical system on theology without sufficiently establishing theology, and the history of that attempt has so far not achieved success. For every proposed proof, there has been at least an equal and effective counterproof.[91]

[90]See Ilya Prigogine on "dissipative structures".

[91]*Readings in the Philosophy of Religion: An Analytic Approach*, Baruch A. Brody, ed., Prentice-Hall, Englewood Cliffs, New Jersey, 1974.

This first move (ground or explanation) is blocked. Unchanging, immediately apparent natural rights that have meaning only within the framework of natural law or a law-giver require either an unchanging order in the universe or a Being who guarantees these rights over and above our human experience. The rights model in medical ethics speaks of universal and unchangeable rights, not modifiable or capable of abrogation (inalienable rights). The first explanations, (1) and (5), cannot meet these conditions.

If the natural order or theological base is missing, an intuitionist explanation can be attempted (Feinberg, Hart, e.g.). Feinberg's understanding of rights as claims ultimately depends:

(1) On legal or social bases, and hence rights could be abrogated, change with social change, and need not be universal.

(2) On an "enlightened conscience" and lack of moral depravity, which is clearly intuition-grounded.

As Macdonald very early pointed out,[92] we do not have intuition consensus, nor have we yet achieved it, nor if we did am I sure what it would prove. I am very uncomfortable with defining an "enlightened conscience" as one that agrees with our rights concepts and then using such a conscience to establish or ground those rights. Something very circular is going on here. Feinberg pays a good deal of attention to the claiming process, but either such claiming is socially relative or it is based on the intuition of a "proper" conscience and begs the question. Feinberg also considers defining a right as a "simple, undefinable, unanalyzable primitive,"[93] which would again prematurely terminate our search for a grounding for rights and finally deprive them of meaning possibilities. He prefers an analysis of the claiming act to such a definition, but since he allows the concept of *valid* claims, we are back to the validation problem which the move to simple primitives (intuitions) solves by allowing the assumption of what needs to be validated. If rights really were simple primitives, difficulties of meaning (such as agreement on lists of rights) would be minor since we would clearly enough understand what we were talking about and use the term appropriately. Nor will it do to label usage of a term that conflicts with ours, or differing intuitions of what rights are or

[92]Margaret Macdonald, "Natural Rights," in *Human Rights* A. I. Meldon, eds., Wadsworth, Belmont, Ca., 1970, pp. 40–60.
[93]Joel Feinberg, "The Nature and Value of Rights," *The Journal of Value Inquiry* **4**, 1970.

are not, as "unenlightened" or "morally depraved." That seems clearly *ad hominem* or circular, and another indication that rights are neither easily agreed upon primitives, nor intuitively apparent claims.

Hart, along with Rawls, who tries to ground it, rather too easily assumes a version of Mill's freedom principle: the right to the most freedom compatible with all others' equal right to such freedom.[94] What may be the foundation for such a right is not discussed, yet it is a complex concept involving difficult notions of human nature both on the individual and social levels. It assumes a good many things about agency, conditioning, social roles, dependency, and autonomy that are not immediately intuited.

McCloskey defines rights in terms of interests.[95] The ability to make claims is essential, and claims are made on the basis of self-perception of interests, or surrogate evaluation of interests. Rights finally, in his analysis, are expressions of interests; but such a conception of rights is again not the one needed for a universal and immutable characterization of rights. The status of such claims is not based on a cognitively perceived necessity, but on the ability to empathize (share interests and understand desires) or feel compassion—to choose to respond to a claim that is really a request. Unless we import illicitly into the argument an assumption of some universal and unchanging human nature that generates universal and unchanging interests, and then assume some motivation to honor claims based on this commonality of interests (both assumptions being debatable), the notion of rights remains ungrounded.

A third move, to talk about "human" rights as apparent on the simple basis of our humanity, our innate worth, our membership in a kingdom of ends, requires either consensual intuiting of this uniqueness and worth, or a demonstration of relevant and important (for ethics) differences between human and nonhuman beings that justifies this unique worthiness, or defining human in a Kantian way (whose problems I will discuss shortly). Rights again become either intuitions or special worth. The intuitions are not universally shared or experienced. Worth can be a circular restatement of rights, as similarly Kant perceived that autonomy ran the risk of being circular for the ethical categorical imperative (the moral law). Kant's solution of splitting phenomena and noumena

[94]H. L. A. Hart, "Are There Any Natural Rights?" *Human Rights*, A. I. Meldon, ed., Wadsworth, Belmont, Ca, 1970.

[95]H. J. McCloskey, "Rights," *Philosophical Quarterly* **15**, 1965.

and giving us a special access to the noumenal realm (a transcendental ego) has theoretical difficulties and certainly viable alternatives. As Regan argued,[96] using this concept of "worth" may involve the following circle: Right to life → every human being is worthwhile → every human being has a right to life. As Beck pointed out with Kant's risk of circularity,[97] if "right" and "worth" are correlative concepts (mutually implicative), then we cannot use the concept of one to establish the reality of the other.

A final explanation, basically a Hobbesian move (it is an interesting aside that in critiquing Rawl's basically social contract move, Harsanyi mentions the contractarian tradition of Locke, Rousseau, and Kant, and ignores the most well-developed social contract theorist, Hobbes, who based his original position choices on affective as well as rational human nature), has been suggested by Murphy and Macklin[98] as derivable from Rawls' *Theory of Justice*.[99] And another interpretation of the "right" and theory-of-justice connection is offered by Dworkin.[100] Murphy and Macklin find rights derived and explained by Rawls' social contract version. Dworkin, however, finds a deep structure beneath this position, which he identifies as the right to equal concern and respect (fairness), an equality principle. Further combined with Rawls' emphasis on a rational equilibrium, this deep structure becomes the ground for specific rights without necessitating a naturalist interpretation (it is a constructionist ground, according to Dworkin). There are two problems with this:

(1) The equality principle is ungrounded (the original position is a mechanism generated by this assumption, not a demonstration of it; it is an expression of the assumption). Our intuitions would not agree in specific situations on the adequacy of this assumption or the choice of it (triage, as an

[96]Tom Regan, "The Moral Basis of Vegetarianism," *The Canadian Journal of Philosophy*, October, 1975.

[97]Lewis White Beck, "Translator's Introduction," *Immanuel Kant: Foundations of the Metaphysics of Morals*, Bobbs-Merrill, Indianapolis, 1959.

[98]Jeffrie G. Murphy, "Rights and Borderline Cases," Kopelman and Coisman, eds., *Rights of Children and Retarded Persons*, Rock Printing Co., Rochester, New York, 1977, pp. 6–20; Ruth Macklin, "Moral Concerns and Appeals to Rights and Duties," *Hastings Center Report* **6**, 31 (1975).

[99]John Rawls, *A Theory of Justice*, Harvard University Press, Cambridge, Mass., 1971.

[100]Ronald Dworkin, *Taking Rights Seriously*, Harvard University Press, Cambridge, Mass., 1977.

example). Fairness is much more complicated than equality—we are fairly, not always equally, concerned.

(2) A constructionist interpretation of rational equilibrium is a coherence theory with certain moral intuitions *not* subject to the coherence compromise, and hence is internally inconsistent, as well as being a disguised form of moral intuitionism. Since Dworkin's "deep structure" interpretation of Rawls becomes an intuition justification of the rights model, my previous criticisms apply here, and I will concentrate on the interpretation that the original position can serve as a grounding for rights.

The decisions made in Rawls' original position (the mechanisms of ideal social contract in addition to two of the old Greek virtues, prudence and the good-as-reason) could be understood as what we mean by "rights" in Murphy's or Macklin's suggestions. My contention, however, is that Rawls' concept of the original position is so untenable that rights could not successfully be grounded by it. Certainly his two principles derived from the original position sound very much like suggested consensual or intuitive basic rights: (1) Each person is to have an equal right to the most extensive total system of equal basic liberties compatible with a similar system of liberty for all, (2) Social and economic inequalities are to be arranged so that they are both: (a) to the greatest benefit of the least advantaged, (b) attached to offices and positions open to all under conditions of fair equality of opportunity.

Like Hobbes, Rawls constructs a hypothetical original social situation that he calls the original position. Although Hobbes' primal situation depends on a realistic assessment of an individual's inability to guarantee personal security, no matter how powerful the individual may be, and a fear of death, Rawls' situation involves a "veil of ignorance" concerning the social role one will have in the negotiated society. Rawls also makes the assumption that these negotiators will all try to minimize their possible losses rather than risk maximizing possible gains (maximin principle) and that this is the only reasonable goal. None of them, then, are risk-takers or passionately committed. Like prudent, rational game theorists (not even as risk-taking as Bayesians) they are all by common nature rationalists rather than affectivists, play-safers and not risk-takers, minimizers (basing decisions on the worst possibilities) rather than probable or all-or-nothing maximizers—Rawls' incomplete rational model of man. I will argue that human nature is not characterized only by logical prudence, that this is a philosophical conceit that makes ethics incomplete, and that furthermore, we would not want human beings to define themselves solely in this way. Some risk-taking is required to be adaptive and survive. No risk-taking is static, no-growth, nonresponsive to the environment. So first, Rawls' version of the agents in the original position is not descriptive of human beings, and second, if the agents behaved in the way he theorizes they would, the results would be bad rather than good, and our "justice" would be self-destructive.

Harsanyi has a similar, but not as extensive, objection to Rawls' theory,[101] based on the advantages of a Bayesian "expected utility maximization" over the maximin principle. Since maximin says you should evaluate any choice available in terms of the worst possibility that can occur making that choice (and that all possibilities except that worst one are nonweighted), it can lead to absurd decisions that would block even simple daily activities like crossing a street. Furthermore, as (2a) states, choices must be made in terms of the interests of the least advantaged (again, so that once more the worst possible *is* weighted, but all other possibilities again are not. Or in Kantian terms, whereas the worst-off individual is treated as an end and not a means only, all other individuals in a choice-conflict with that individual are treated solely as a means to an end.

Harsanyi considers and eliminates possible objections to this critique:

(1) Counterexamples may not prove much (Rawls). Although true for examples involving minor points in a theory, this is hardly true for the fundamental mechanisms of a theory.

(2) Subjective probabilities should not be used in the original position (Rawls). Rawls wishes to eliminate not only empirical probabilities with his veil of ignorance, but subjective ones as well, and, illicitly, all gambling behavior. Gamblers, however, do not *require* empirical probabilities, or even subjective *good* odds, but make a choice of risk based on evaluating the *desirability* of the goal over the undesirability of doing nothing, and can only be confined to maximin (or even Bayesian) by fiat. Furthermore, as Harsanyi points out,

---

[101]John C. Harsanyi, *Essays on Ethics, Social Behavior, and Scientific Explanation*, Kluwer, Boston, 1976, esp. chap. IV: Can the Maximin Principle Serve as a Basis for Morality? A Critique of John Rawls' Theory.

maximin implies a subjective probability assignment of one or nearly one to the worst possibility in a given case.

(3) von Neumann-Morgenstern utility functions have no place in ethics (Rawls). This is merely another statement of Rawls' assumption of nongambling behavior and of prudence as the only virtue. To define all except nonrisk-taking and prudence as nonmoral is to assume what one is supposed to demonstrate and builds into his system exactly what he intends to get out of it.

(4) Maximin is a macro- and not a micro-principle, therefore Harsanyi's objections (micro) do not apply (Rawls). Harsanyi very easily points out that his objections will work on the macro-level just as effectively (allocation of medical personnel and resources according to triage decisions or according to the worst-off principle, those hopelessly ill, e.g.). Although Harsanyi is puzzled over what difference levels of organization should make in ethics, and I on the other hand find such levels an important analytic tool, in this case Harsanyi's answer is more than adequate since he can apply his objection to the macro-level with full force.

Therefore, the social contract model does not accurately describe the human condition and even if we argued that human beings should operate with only the rational prudence it dictates (which actually has disvalue in terms of survival and system stability), there are better or at least equally attractive alternatives to maximin. Secondly, even if we waived all these major objections, unless we assume an invariant human nature as described by Rawls, the theory will not give us the sort of absolutism (universal, immutable, eternal) that the rights model requires. Our "rights" would be contingent on what particular aspect of evolving human nature happened to be emphasized, and would therefore vary and change.

We do have legal and social consensus rights, of course, which is what confuses things. These are rights given to us by society's legal expression, things we give ourselves, actually. They are guaranteed by the social system in codified ways, with enforcement provisions. Social consensus rights are part of the mores of a society, agreed upon, but not written into the law, functioning as norms of human interactions, expected behaviors. They are also culturally dependent and vary as the social system adapts and changes. Major social system dislocations can eliminate them, as the law's interpretation and evolution alters or eliminates some legal rights. So legal and social consensus rights are given, and what is given can be taken away. That is *not* the sort of right a rights model requires. The model's rights are not humanly given and cannot be taken away: they are immutable, eternal, and universal. I hope I have indicated that, from a critical philosophic persective that asks for grounding or justification, they are also non-existent.

There is one last danger in using this model. A rights model is usually embedded in a legalistic form, which reflects its old natural law beginnings.[102] However, constructing all human relationships in a legalistic framework is a serious distortion of human interactions, especially those where affect, bonding, or complex interactive processes are involved. In addition, such a "by-the-book" approach assumes that we can predict all possible situations, can codify all needed solutions, and can fit variety and change into an apparatus that will not be so cumbersome as to collapse of its own weight. It assumes we can program all human interactions, present and future. Although there is a cultural bias in favor of such legalism, I want to suggest we are a society of human beings, not of rules; that rules derive from our existence as social beings; that a social system is richer than rule-following or it could not survive; that even the legal role is not as legalistic as this model assumes (most routine decisions are worked out through personal interactions before the parties get to the bench, e.g., plea bargaining, jury selection).[103]

These five proposed justifications for a rights model equally fail then, and it would make an interesting study in sociology of knowledge to inquire why they are so easily accepted and why the legal metaphor is such a commonly used one. I gave a practical example of the failure of the rights model in Chapter 5. I can see no way, except by translating it into a legal rights model, to theoretically establish a surrogate role using it. How can "rights" be transferred? By what mechanism is a child's "right" given to a stand-in who then decides whether to exercise it? In actuality, we know that social bonding is the social base for the legal surrogate role. Its actual establishment is based on socialization and affect, not on a mysterious unit that can also be mysteriously implanted

---

[102]Contrary to Veatch, the rights model is not so easily disengaged from the legal model. See Robert M. Veatch, *Case Studies in Medical Ethics*, Harvard University Press, Cambridge, Mass., 1977, pp. 86–88.

[103]For another critique of Legalism in Medical Ethics, see John Ladd, "Legalism and Medical Ethics,"*Contemporary Issues in Biomedical Ethics*, J. Davis, B. Hoffmaster, and S. Shorten, eds., Humana Press, Clifton, New Jersey, 1978, pp. 1–35; and Alasdair MacIntyre, "What Has Ethics to Learn from Medical Ethics," *Philosophic Exchange* 2 (4), 37 (1978).

in another individual. It is a short conceptual step from "rights" and theories of justice that incorporate Mill's liberty principle to the notion of moral autonomy as an overriding value or the prerequisite for morals. An emphasis on moral autonomy (from Kant to Wolff) brings us to the second common system in medical ethics.

(2) Person as End/Respect
for Persons

Respect for Persons can take two forms, either a Kantian person-as-ends or a general humans-as-special status. The Kantian base—depending on his philosophic system's constructs of kingdom of ends, humans as importantly an expression of reason, and reason as transcendent—could more accurately be described as respect for reason (and finally, noncontradiction). Basically, it would be a contradiction to treat humanity as an object only, since we are all rational beings obeying a common law (of which one important principle is the principle of noncontradiction), and therefore members of a kingdom of ends, as well as objects. To treat another as an object only is to will that we ourselves not be ends, but since we embody reason, we are ends, which is a contradiction (we must choose not to commit). "Respect for persons" is an odd way of phrasing this in the medical ethics literature, since "person" and "reason" get ethically equated and reason becomes our relevant (and ethically *only* relevant) definition. Unfortunately, we can always question valuing or accepting noncontradiction as overriding. As Freud pointed out, descriptively, our primary processes frequently function as contradiction, from a formal viewpoint. We also know that human beings are defined relevantly (if we are not begging the question by using "relevant" to define away other human potentials) in many more ways than rational and there seems no clear reason to prefer rational over other definitions, e.g., affective. The justification again for Kant is his phenomena/noumena distinction or dichotomy, with really an undeveloped assumption of higher value to the noumenal (and humankind's rational ego as part of it). Our affective and physical egos seem to enjoy less status even than the Greek troika gave them, but certainly repeat that value ranking of humanity's various "natures."

On more practical grounds, even given Kant's ethical emphasis on avoiding contradictions, it is not that easy to define or demonstrate a contradiction. Whether there is a contradiction depends on interpretations that can be variable or assumptive. With sufficient interpretation of the premises, we could conceivable have all contradictions and no maxims, or no contradictions and anything a maxim (suicide, e.g., can be seen as either). This is not a completely technical point, but demonstrates a weakness in one of the fundamentals of Kantian ethics: the use of formal reason as a ground for an ethical system.

What happens in medical ethics as a result is that the actual Kantian program for morals is really not applied or cannot carry the burden of what is really intended by respect, and through unstated meaning shifts, we really talk about norms of well-functioning (respect for potential worth, self-regarding), agapé (Christian brotherhood), or empathy (respect as another human being)—that are not made explicit.

Another theoretical basis for person-as-end is a once theologically grounded separation of man from animal species. Humankind was made in God's image (theological) and alone had a soul. Humans had reason or mental aspects that made them unlike other animals (Kant, Cartesians, the traditional philosophical separation). Humans also had feelings, unlike other animals (emotivist), and had language (the modern philosophical separation). One by one these gulfs have been bridged. Philosophically, before we can appeal to the ideas of soul or of being made in God's image, we have to establish theology. Since for every rational proof, there is a counterproof proposed, we may have faith that human beings are special, but not yet a philosophically established reason for that belief. Nor do all theologies, in fact, preclude animal souls, so that we would need to establish a very specific theology in which animals had no souls. Reason lost its distinguishing ability when animal learning was demonstrated and when higher forms of animals were shown to engage in problem-solving. The primates certainly share some of our mental category development, following a Piaget model, as experimentally demonstrated. Emotions and expressions of social bonding are common to many social species besides ourselves, as the ethologists have clarified. The signing primates, in addition, sign feelings they have (sadness, separation anxiety, shame or guilt, anger). We have finally fallen back to the point that language is our key line of separation, but that battle is really lost as well. The signing primates, using Ameslan, demonstrate that either animals can communicate or that deaf and mute individuals are not human beings. (I am being facetious, of course.) Ameslan is an acceptable language capable of artistic communication as well as everyday functions. The primates exhibit construction, syntax, and generalization (universals).

We are, then, special to each other, and our interests and needs are special to us, but we are not so separate from the rest of the biosphere. "Special to each other," however, is just another phrase for a human affect I have been stressing in ethics: empathy. So·the respect model becomes either a systems theory ethics of well-functioning or an emotivist ethics based on empathy, which is very different from the formal and deontological system its proponents really want.

There is one last *a priori* possibility offered as justification for some respect model, the use of moral autonomy as the *a priori* condition for moral obligation itself. The argument would go: Moral autonomy cannot be given up on moral grounds since such a decision or choice would be itself an act of moral autonomy; but not to give up moral autonomy is to continue to be morally autonomous.[104] If our perception of morally autonomous human beings results in what we mean by "respect" for them, then this can be offered as a ground for respect *and* for the overriding value of "autonomy" in ethics.

However, to make this argument work, there are some assumptions packed into it that we would need to accept:

(1) Ethics involves ultimately *only* subjective good, so that the important concern in ethics is for the internal state of the actor, as the actor perceives it.

(2) Autonomy is the choosing behavior of an agent motivated by rational considerations most importantly, not influenced by social bonding, but seen as discretely asocial, and also not relevantly influenced or motivated by affective elements.

(3) Decision-making occurs at a definite point in time, rather than being an ambiguous evolving process.

(4) A decision can be made at only one level; there can be no meta-decisions. Also, it is assumed it is legitimate to construct a self-referential statement concerning choosing to give up moral autonomy.

None of these assumptions are obviously true: (2) and (3) are contrary to fact, descriptively incorrect; (4) is not only contrary to experience, but there is a body of literature to suggest that one should *not* construct self-referential statements, that these are linguistically unsound (Tarski, e.g.). Furthermore, systems theory identifies many levels of attention, discourse, interest; and rela-

[104]Jeffrey H. Reiman, *In Defense of Political Philosophy: A Reply to Robert Paul Wolff's in Defense of Anarchism*, Harper and Row, New York, 1972.

tionships are different at these various levels, so that meta-decisions concerning moral autonomy need not conflict with the subject matter of those decisions (which might be "moral autonomy" at the lower level). On quite moral grounds, knowing I am a psychotic killer, I may surrender my moral autonomy, and the decision process makes perfectly good ethical sense. One could argue that autonomy should only be surrendered in choices concerning the life or safety of other individuals, but if I were unsure of my ability to distinguish those situations from others, I might very well and morally make the decision to give up all moral autonomy. Statement (1) represents only one possible position in ethics and would lead to conclusions such as: a drug-induced, euphoric state that is physically destructive is nevertheless good.

Again, the practical failure of the respect for persons model to justify surrogates is based on difficulties with the application of formal structures for contradiction, so that there is a real possibility that any maxim can be a universal law of nature and that all of us, as surrogates, will not reach agreement or reach contradictory maxims. Furthermore, the surrogate status itself is based on an odd notion that humans are most importantly rational and that "end" is equated with "reason," and because of that is transcendent. Actually, although it is not quite the other way round, the reverse would be closer to the truth of the matter. An end is a goal or purpose, and although reason was certainly an important purpose for Kant, it is only one of a variety of human purposes and often a component in a larger interest.

Attention to objective as well as subjective markers for the good is one of the possibilities of the third ethical system common to medical ethics, where definitions of pleasure involving either a subjective state, an objective state of well-being, or both are alternative suggestions.

(3) Utilitarian Consequentialism

Mill proposed utilitarianism as an alternative hedonism and in medical ethics one form, cost/benefit, is extensively used. The good was pleasure (from happiness to objective well-being), but a calculation (another significant problem) had to be done to achieve the greatest good for the greatest number. This looks as though it lends itself nicely to macro-ethics (social interest ethics), but may run aground in a macro/micro-ethics conflict, prejudging in favor of social interests. My main criticism of utilitarianism, in fact, is based on systems theory; namely, that it confuses two distinct levels (social good and individual good), and is not a traditional critique (of which there is ample literature; see Brock's survey[105]). Not only are levels illicitly mixed here, but I hope to show that a balance is required between social good and individual good forces to keep a social system intact, and that a utilitarian ethical system cannot meet that requirement. It has built-in mechanisms for sacrificing individual interests to social interests (by equating or overriding), overlooking the reality that a sufficient number of unmet individual interests will cause extreme enough disruption within a social system to result in collapse. These unmet interests need not outnumber the aggregate good to result in this conclusion, so that a calculus will not save the utilitarian argument.

Nor does utilitarianism theoretically consider the inescapable conflict involved in ethics, when it considers only the aggregate good by redefining individual good as social or common good, involving it in a calculus that gives it relevance only in terms of numbers. It mixes individual and social interests in the same calculation, treating them as if they involved the same level of analysis. This is a different problem from that of interpersonal utility comparisons, which are actually done at the same level (Harsanyi's "basic psychological laws" solution may or may not work, but it functions on the individual level). This problem involves coming to ethical decisions at a social level of interest, with the individual level relevant only in aggregate (hence, social) terms. A single individual is therefore always a potential sacrifice to the aggregate, which explains Brock's eagerness to introduce the mechanism of justice or fairness into our ethical considerations. The lack of a mechanism to provide a power base, if you will, for individual *qua* individual interests in utilitarian theory makes it a seriously oversimplified theory of the nature of social and individual relationships. I have here given a systems analysis of this relationship that should provide an understanding of the basic flaw in Utilitarianism.

The present discussion has not touched traditional problems with this ethical system in a comprehensive way for two reasons: I think many of these objections can be met, and the objection I propose is so basic that utilitarianism will stand or fall on it. Generally, the traditional criticisms fall into these categories: problems of comparing pleasures or degrees of pleasures or even defining pleasure; problems of choosing between subjective well-being and objective well-being; problems of justifying minimization for a large number over maximization for a few; problems of interper-

[105]Dan W. Brock, "Recent Work in Utilitarianism," *American Philosophical Quarterly* **10** (4), 241 (1973).

sonal comparison; problems of determining consequences; problems with an intrinsic/consequentialist dichotomy; problems with act vs rule or ideal rule utilitarianism, and utilitarian generalization; problems with the conflict between utilitarian decisions and ordinary moral convictions (the demand for congruence), and especially discrepancies with fairness or justice perceptions; problems with punishment and excuses; and the problematic status of "rights." Finally, since this ethical system can reduce to trying to achieve the least amount of pain for the greatest number, how does one answer Silenus' objection that it is best not to be born at all, and next best to die as quickly as possible? Or Aristotle's observation that you could never tell whether someone had lived a happy life until it was over?

For the cost/benefit version of utilitarianism, the specific problem is quantifying benefits that involve soft data. The solution is usually to exclude soft data and concentrate on hardware and billing. This distorts the ethical analysis so seriously, by removing the human element that interests us in the first place, that it is a false calculus. In addition, cost/benefit analysis has trouble quantifying individual need or desire, and usually defines it by the marker "how much willing to pay." It often makes assumptive protocol decisions about what will be considered consequences. And finally, it has the utilitarian mechanism for sacrificing individual interests to social interests.

In my practical example of surrogate role, the utilitarian model works even more poorly than the other two. With the perspective of the common good, there really is no meaningful surrogate role within this theory, since that role implies the perspective of the individual good. The model has built-in mechanisms for always sacrificing individual needs and interests to the common good, which is precisely what the surrogate role is designed to prevent. The confusion of social interest and individual interest levels is very apparent here as a large flaw in the model.

When a cost/benefit calculus is considered, not as a social interest form of utilitarianism, but as a tool for determining individual interests, it is really consequentialism. I will argue that all ethics is ultimately that, and that deontology is a pre-determined or closed consequentialism. We make the "right" choice because we already believe that such right action is so in tune with an assumed order of the universe that it is always a good result without having to bother to do the calculation. We intend the right thing because we already think we know that such a choice always has a good result (another way of saying—and spelling out meaningfully—"is a good"). We made up our minds about it and will no longer calculate. To the extent that rule-utilitarianism does not include an analysis of justification of the rules, or the etiology of those rules, it differs only by stressing the results rather than the intentions and psychological results to the intender. But intentions are important only in terms of responsibility or future psychodynamic *results* to the actor, and represent an incomplete analysis of the situation, although still consequentialist if incomplete.

Ethics involves a consequentialist cost/benefit calculation, then, keeping in mind that psychodynamic results to the chooser or actor are one of the costs. But it involves much more this, more than adaptation and more than an application of the pragmatic criterion. Underlying consequentialism, which is ultimately based on a pleasure/pain principle, is the assumption that acting to produce bad results will somehow pain us, give us an unpleasant feeling. If it did not, if the capacity for this affect were not present, ethics would not be possible. Philosophy's overlooking this requirement for ethics explains many of the traditional criticisms of proposed ethical systems, e.g., the criticisms of egoistic hedonism as unworkable. However, the critics tend to forget that the same flaw is built into their own systems, which also are cognitivist rather than emotivist, and neglect affect as well.

Pragmatics

Exactly what does ethics involve, then? I have indicated that the model traditionally used is overly rationalistic. In addition, philosophic love for dichotomies, either/or binds, has artificially separated valuation from a supposedly interest- and purpose-free world of facts, severing feedback loops from experience. The uncritical acceptance of an is/ought distinction, in fact, is a regressive move by ethics back to the old epistemological subject/object problem. A pragmatic ethical system needs to avoid both these fundamental flaws.

To the extent that a naturalistic ethics was categorical, it reflected an inaccurate, static view of experience. More than a descriptive tally of ethical choices, such an ethical program attempted to find natural characteristics of humans and their interactions with the environment that were the ground or deep structure for subsequent choosings (valuings). The clauses, however, for such a program would have to read: "you wish to avoid pain, so ____", rather than, "*if* you wish to avoid pain, then ____." Such a formulation assumes an invariant environment, an unchanging and consistent human nature, and a pre-set and immutable relationship between the two, which it is also assumed equals well-functioning. This latter assumption is significant. It is accurate only if well-functioning is defined in terms of survival and *if* survival is valued. This is the catch for traditional naturalistic ethics. More attention to the primacy of affect in all valuation would avoid this error. To the potentially changing nature of man transacting with a potentially changing environment (which already importantly alters such an ethical system), we need to add a basic and also potentially changing affect response of affirmation or negation. This addition makes a categorical imperative (rule) impossible in ethics, but it most accurately reflects the stasis involved in our human natures and the evolutionary process in our transactions with experience.

For any valuation to occur, as I have argued, an affirmative attitude is required. Choosing, being concerned, having outcomes matter, implies it. So our fundamental imperative (rule, generalization) is hypothetical: If one affirms existence (experience), then that individual will try to maintain functional norms (in old or new experimental, evolving ways), and then will make those choices that to the best of our knowledge will realize this goal.

All valuing, then, depends on an affirmative feeling about existence, which is a condition for any value system (cognitive, esthetic, or ethical).

There is another affect that must be operative before ethics is possible. Unless other people make a difference to us (empathy, social bonding), ethics is not possible. That is why, for example, for ethical egoism to work (and it will), we need to assume a socialized human being, one capable of satisfying object-relations, who recognizes that it is truly in his or her own self-interest to express social characteristics. The same is true, however, for utilitarianism. An individual values the greatest good for the greatest number because that individual is a socialized human being, and can empathize. The notion of justice (or fairness) is a socialized individual's concept. The rights model, in any of its forms, if pushed to ultimate justification, involves social characteristics and the ability to empathize or it will founder. Only the Kantian ethic completely replaces empathy with an asocial value of noncontradiction. But this model never asked why noncontradiction was valued, or why universalizability (consistency) was valued. The answer to the latter would finally imply a socialized human being. The answer to the former would ultimately hinge on survival, and thus affirmation of existence. Failure to appreciate affects as a crucial part of ethics again explains why social bonding and empathy remain the missing key for meaningful, grounded, and workable systems.

Ethical values depend on our socializing affect called empathy. Empathy is an expression of our social nature, but although it is a biological predisposition, it is a developmental process that can fail or be seriously defective.

Ethics needs to be extended to include this arational component or distortion and conceptual confusion will result. Empirical psychology can be of considerable assistance in this task by filling in the missing affective parts in the system.

If we affirm existence and are sufficiently socialized, we can construct a pragmatic criterion (based on well-functioning, survival, opportunity for pleasure). The application of that criterion can be empirical (objective), but the entire generalization is hypothetical: If we affirm existence as socialized human beings, then in this particular situation, with help of rules of thumb from previous generalized experience, in order to achieve our purpose of well-functioning, we should_____.

There is a last attitude or affect that, specifically in medical ethics, needs to be worked into the system, but I will also argue that since ethics is best seen as choosing (valuing), needs also to be a part of a general ethical system as well. All along, I have presented various forms of an ethical justification for intervention, with medicine as the exemplary interventionist human behavior. I have also elaborated two basic attitudes toward pain and suffering: intervention and acceptance. Do we wish to change certain features of our environment or our relationship to it, to change certain features of our future projects? Or are we willing to accept things as they are, and to resign or adapt ourselves to what will be? Medicine, as interventionist, tries to avoid pain by changing circumstances, by making a difference in projects, by *choosing* rather than modifying internal responses. Of course, we can always choose not to choose; that too is a choice, as the existentialists perceived, but it is a final choice, closing out and not opening up options. Ethical systems are not characterized by that finality of choosing or valuing, but by a continuing choosing and valuing. They do not have to be, and we could construct systems with such finality since it is an attitude, an affect that we certainly are capable of expressing. In many cases it may not be survival-promoting, and can be faulted on that basis, as on its static human/environment relationship, which bodes badly for adaptation and continuing stasis. The critical point is that the problems ethics is constructed to handle involve attempts to choose the better option because the choice does make a difference in our relating to experience. If we were not concerned about the differences, if in our equanimity nothing that occurred mattered at all (that stoical goal), if choices beyond the acceptance choice really did not concern us, then ethics as such choosing and valuing would not concern us, would not be an activity for us. I hope I have demonstrated that ethics makes sense in terms of human choices, that valuing is choosing, and that there is very little sense in any other terms. If that is true, then ethics is at base an explanation and justification of interventionist behavior and affect. Medicine is applied ethics. All through the medical enterprise runs the thread of interventionist choices, their analysis, and their justification.

To summarize the system:

(1) All valuing depends on an affirmative feeling about existence, which is a condition for any value system (cognitive, esthetic, or ethical).

(2) If we affirm existence, we can construct a pragmatic criterion (based on well-functioning, survival, working well, cash value, opportunity for pleasure, and so on).

(3) The application of that criterion can be objective, in an empirical sense, but the entire value statement is hypothetical: if we affirm existence, then (in this particular situation, with help of rules of thumb from previous generalized experience)____.

(4). In addition, ethical values depend on our socializing affect called empathy. Since we are a social species, such empathy or compassion is a biological predisposition. In fact, studies indicate it is essential (as social bonding) for our physical survival even when other physiological needs are met. Unless other people make a difference to us, ethics is meaningless. But they make a difference because of our affect, a feeling of empathy.

(5) Specifically, in medical ethics, another affect is crucial, our attitude toward pain and suffering. Are we interventionists or accepters? Do we wish to change certain features in our environment or in our future projects, or are we willing to accept things as they are and resign or adapt ourselves to what will be? Do we want to avoid pain by trying to change our circumstances?

Value of a Systems Model

The approach developed throughout this entire volume assumes there is value in adopting a new scientific paradigm—systems theory. However, discussion of the advantages offered by this approach have been scattered through the various chapters, and I should like here to attempt to draw them all together. Many traditional problems can be resolved using this new approach, e.g., mind/body dualism, reductionist/emergent properties mechanism, social/individual interests analysis, the definition of "persons", identity as structure or material, and so on. However, two basic systems concepts carry most of the load, specifically, those of stasis and hierarchy levels.

Not everything called a systems approach is founded in general systems theory, and the stasis concept makes the necessary distinction. Management design systems are sometimes confused with the systems theory model, and it is best to clear that up in the following. There are some common elements that add to the superficial confusion. Management designs frequently use line and staff arrangements and intricate flow charts. There may even be some feedback loops, and at first glance these appear grounded in systems terminology. However, line and staff are designs for command flow and are frequently unidirectional. Negative feedback loops to maintain the system's balance or stasis either are not present or cannot significantly alter the function or form of the most important, directional components. It is not an evolving structure, but a chain of command. Flow charts primarily move the product (information or material) along predetermined pathways that again are not much influenced by negative feedback mechanisms from the subsystem components. Internal oscillations and corrections of structure or function are rare. The product itself has little feedback into the system. Finally, where feedback loops are present, they tend to connect from the end of the chain of command to the top component in the design, or the end of the flow, are not particularly elaborate, and do not entail intricate interconnections with other components. This is very much unlike the homeostasis of the human body, for example. Another instructive example of this management rather than general systems model is Dr. Frederick Glaser's Systems Treatment for Alcoholism program at Toronto.[106] Here the Assessment Component (the real decision

[106]Frederick B. Glaser, Stephanie W. Greenberg, and Morris Barrett, *A Systems Approach to Alcohol Treatment*, Addiction Research Foundation, Toronto, 1978.

area) has a command line to the Treatment Component and the Treatment Component's data are interpreted by the Research Component, which feeds back up the line of command directly to Assessment. Research can modify Assessment, which again unidirectionally controls Treatment. Except for the two-way feedback between Assessment and Research, the total design has separate job descriptions with the Treatment Component's feedback severely limited. Rather than continuing to provide for change, oscillation, or adaptation, the Assessment structural goal is to more rigidly predetermine Treatment choices and make increasingly fewer exceptions to individual assignment. It is assumed, in fact, that this rigid structure is not only desirable, but also possible. There is little opportunity in the system for new configurations, for adaptation and continuing growth, all of which are necessary for systems maintenance.

Homeostasis is a more complex notion than this, an exquisite balancing act that allows form the changing structure necessary to maintain itself with controlled internal oscillations. Such oscillations become the material for adaptive change, and are not nuisances to be eliminated, but the building blocks of adaptation over long periods of time; systems are intrinsically adaptive.

Hierarchy theory, by identifying levels of organization or interest can explain the complex structure within the framework of matter and give reductionism the necessary theoretical connection with changing properties and "entities." Identifying the terms and conceptual tools of different levels of complexity allows us to avoid mixing these metaphors in quite a real sense. The thoroughly confused discussion of abortion in the current press illustrates both the conceptual and actual practical costs of such confusing mixes. Conceptual difficulties with social level or individual interactive level terms like "person," "purpose," "interest," or "act" can be resolved within a hierarchical framework, provided that the levels of capacity are not viewed as exclusive ontological categories, but as complex structures sharing the common building material. It is often pragmatically useful to construct epistemological entities that depend on our purposes and interests (and also on the possible structuring of existence). These are not mysteriously different materials from those of the more reductive levels, however. We could in theory restate the terms of a more complex level at a more simple level, but the unwieldiness of the translation would not make it worth our while for most purposes, and might make it a practical impossibility in terms of the components and relationships to manipulate.

Such systems-level considerations will explain emergent properties, even general creativity, not as some awesome and non-understandable event, but as the process of changing configurations and increasing levels of complexity. It will certainly make clear what we mean by identity, or have us make it clear: are we focused on material, or identity of structure? It will certainly handle the mind/body problem, in the same way that it handles the "person" concept, by requiring us to pay attention to our different purposes, uses, and foci, when we employ these terms. Finally, as our last chapter demonstrated, it will really allow us to understand the social ethics/individual ethics conflict in a way that forces us to recognize the hard choices and tension that are inextricably a part of the problem. Our ethics will not become unrealistically utopian. The systems mechanisms for avoiding internal and external collapse are not always compatible, and in an unavoidable conflict, there is no cost-free solution. Ethics becomes an attempt to maintain a balance, a working tension, between the mechanisms, since overemphasis on one over the other will result in what is meant to be avoided, systems breakdown.

There are good reasons why physicians should not institutionalize an alliance with social concern mechanisms. Rather, a social concern role needs to be applied ad hoc. The patient is already playing a power game with social forces, and in a period of no-growth, resource depletion, that power base is already seriously eroded. If the physician, through an institutionalized social concern role, joins with the social interest mechanisms, the individual patient will be badly compromised and the resulting change in the physician/patient relationship would have serious consequences for the practice of medicine. Here there may be a genuine clinician/academic medicine role conflict. Academic medicine tends to be concerned with the relationship of the medical subsystem to the other subsytems and to the entire social system, to the well-being and growth of the medical enterprise. It is entrusted with the continuing transmission of medical science, its welfare in the larger system, and its development within that system. Clinicians are concerned with the actual practice of medicine, and although as part of the enterprise and establishment they are certainly interested in medicine's social activity, their primary concern is individual practice, which results in quite a different perspective.

The conclusion, using a systems model, is that "dilemmas," that "tragic choices" need to be taken very seriously indeed and are not only emotive figures of speech. We need to avoid expecting too much of ethics, or developing a too-powerful, too-benign sense of what human options involve. The organization of complex systems itself implies conflict of needs and interests. It is more than the tragedy of the commons, which is, after all, only an alternative of economic and political organization. It is the human condition, in the systems metaphor.

This also has implications for concepts such as justice and fairness, rendering their general meanings very problematic. The best sense that may arise from a systems model is my suggested goal of balance (a theoretical base for historical checks and balances). This substitutes core meanings such as determining necessity of need or sacrifice of interest (unavoidability of conflict), buffering of effects (compensation), adaptation of systems goals under force of internal pressure (institutional change), and compromise for the unrealistic meanings attached to these terms (equity, logical ranking unaffected by human purposes, proper and satisfactory solution). I do not want to suggest that balance is an overriding value, however, only that it is a value built into systems maintenance. In many situations, systems breakdown rather than maintenance may reflect the least painful choice or the one giving an opportunity for the most happiness. Opting for new configurations, new structures, is at least an equal ethical stance. Risk-taking is a justified response in terms of growth and creative transaction with experience. Contrary to the traditional ethical models, prudence is not the queen of virtues; we do not have to opt for minimization of losses, for the static response of playing safe. If we do, we buy such prudential peace at the cost of preprogrammed transactions with changing experience, a move that narrows the possibilities of human intervention and thus narrows the chances of adaptation. In a systems view, then, there is an intimate connection between risk-taking and intervention, and risk-taking has considerable value in its own right, rather than the disvalue of its traditional role, which inadequately explained its place in human life.

The philosophic notion of fairness has always been a rationalist concept, not a description of actual experience. Although we have always known that life is not fair, until the systems model, the theoretical underpinning for that practical knowledge was weak. We continued to try to function with a concept that had as its goal the removal of human purpose and interest from choosing or deciding, and often degenerated into the removal of human intervention, of any choosing whatsoever (as in a lottery or predetermined legalistic responses). Such unclear concepts, without feedback loops from experience, from actual situations, frequently caused inhumane results: the equity of rights between fetus and mother requiring non-interference in the death of the mother; the inability to deal with "lifeboat" situations where only a no-cost equity of interests solution is mandated even at the consequence of all interests being lost; the letter of the law choice when the spirit of the law (human interests and purposes) is our affective response; the loss of compassion in the maze of regulation.

A systems approach also indicates why rules of thumb (pragmatic generalizations) will work where rules of practice do not reflect experience. Using the systems concept of stasis, we incorporate change into our description of experience, as well as pay attention to process mechanisms. With this description of our environment and our continuing relations within it, the problem of induction becomes clearly a part of ethics. Generalizations tell us something about past experience, but only in a hypothetical, pragmatic way. If we assume that present and future experiencing will be best handled by unchanging rules from past experience, then we will have systems theoretical evidence that this is a mistake, and the general problem with induction will confirm that. The only method of avoiding that would be to adopt the approach that philosophy strives for in ethics: the separation of "ought" from "is," the placement of values in a transcendent realm beyond human choosing or experience. But I hope I have demonstrated that valuing has as its foundation the human affect of affirmation, which grounds it firmly in experience, in human experience, and supplies a naturalistic, if hypothetical, base for "ought." "Ought" is not based on some rationalist law of experience (which makes no sense divorced from its theological pinnings), but on how we *feel* about experience, and the feeling is a relationship within experience.

Finally, systems theory can give us a view of the end/means, intrinsic/instrumental issue. These distinctions are determined by our purposes and interests, which result in varying perspectives. Both means and instrumental values are such because we place them in a context of broad goals, make them components of a larger purpose, and characterize them, in that context, with the value we place on the total purpose. As long as we are talking about this valuing complex, this level of interest, means and instrumental values are subsystems that need to be considered only in terms of the interrelationships of that interest level. The means are part of the ends. All value is intrinsic until we switch levels.

The Nature and Justification
of Intervention

I have indicated that medical intervention can be either physiological or psychological, depending on our level of interest and purpose, and that the major characteristics and the ethical justification are not much different at either level. Intervention can be necessary or unnecesary, invasive or non-invasive, showing immediately apparent consequences or not, part of a cooperative decision process or of a unilateral one, but it always entails an election to interact with experience in ways humanly chosen to alter that experience for our purposes and interests.

Informing, as I discussed in Chapter 2, is intervention and can be very powerful intervention indeed. Whether it has value or disvalue depends on its contribution to survival and well-functioning. We can usually assume that reality testing has value, and that intervening to produce such reality testing is the right choice. But we assume it, not on some principle of truth-for-truth's sake, but on the basis of a human interaction between physician and patient to determine whether informing will probably lead to such a good (well-functioning). We can generalize to produce a rule of thumb that to facilitate or force reality testing is usually a better choice than collaborating in denial, and our evolving nature has gambled on this, but it will not always be an applicable rule or generalization. Since an informing intervention raises anxiety, and since anxiety can either motivate change and problem solving or create a block resulting in inaction, we have to try, through the physician/patient interactive process, to determine which of these is likely to happen, as well as the real possibilities for problem solution. In some extreme cases, denial can be legitimized, but only if the hopelessness of the situation is accurately determined and if even social support and caring will not be sufficient. In other words, if intervention will not appreciably change the situation, and if the limits of human adaptation will be exceeded, then information is a disvalue and informing is on balance pain-producing (cruel) behavior. In general, however, raising anxiety is not sufficient argument against intervention, since anxiety is a necessary feedback for maintaining stasis, for physical survival, and for any social relating.

Specifically in medical ethics, however, this means that although informed consent is usually a value and one frequent interaction in the relationship between doctor and patient, it cannot assumptively override denial needs. Nor can it assumptively override the expert role, which is a crucial part of the relationship. The expert does not have to collaborate, at least ethically, in the patient's folly. This also means that autonomy is not the ultimate justification for (or against) intervention, since autonomy is often used in the service of self-destructive behavior and we do intervene in self-destructive behavior. Can such intervention be justified? I think my developed justification avoids the rights-model/respect-for-persons model conflict on this issue, as well as clearly spelling out the respect model's unstated dependency on notions of well-functioning or norm:

(1) There is no philosophic reason to value self-preserving over self-destroying, since all values (including the cognitive ones philosophy is based on) are part of, or generated by, a self-preserving affect.

(2) The decision to intervene or not in self-destructive behavior is made by those within a self-preserving mode. They can decide that damage to themselves overrides compassion or empathy for the person exhibiting the behavior, even empathy for autonomy needs.

(3) Social roles generalize and transmit those decisions, with the understanding that they must be reasonable, accurate generalizations and cannot be assumed to handle every present or future situation.

(4) Since such roles, in a changing experience, can become maladaptive over time, feedback from the practice of those roles is required. Intervention is not forever justified.

Research, whether therapeutic or nontherapeutic, also implies intervention and its justification. Research itself is clearly an interventionist attitude toward existence whose purpose is to alter present or future circumstances. It is not based on secondary considerations of consent or covenant, but on (1) the ethical justification of intervention, (2) natural and arational expressions of empathy, and (3) the social nature of human beings, which creates and requires channels for the expression of caring and bonding behavior. Again, there is the question of level of intervention, of how invasive the researcher should be. The real issue seems to be useless, meaningless risk, which deprives the subject of the social affect benefits of participating, and by making intervention useless, deprives both subject and researcher of the good consequences of intervention attempts. The trying is futile and meaningless. The uselessness of many natural experiments (purposeless risk) is the most damaging thing to be said about them. The risk and pain are all for nothing. Certainly the ethics of natural experimentation is a major issue in medical ethics that has barely yet surfaced.

Finally, abortion depends on our justifying intervention in the pregnancy process by terminating it. It further involves viewing such interventionist choices in a way that avoids saying something negative about ourselves, both our worth and our self image. And last, it necessitates legitimizing killing as an ethically justified way of intervening. Systems theory provides considerable help here in avoiding the confusion this analysis is so often subject to. By identifying the different levels of complexity or structuring involved in destroying, killing, homicide, and murder, and by placing those interventionist choices within the biosystem, it helps us avoid muddling our meanings and also, very importantly, points out the human condition.

The Forgotten Human Condition

There are frequent instances in the literature of medical ethics where the human condition is ignored, replaced by an idealized unreality, or distorted by a concentration on only one aspect of it. My discussion of avoidance, of honestly coming to grips with killing (homicide and murder), is illustrative of at least the first two. Because of its existential threat, killing-as-necessary-for-existence is frequently disguised. Yet the killing of some living structures is built into the biosystem. We kill plants to survive, and to the extent that eating is a pleasure (a good), killing can even have traditional intrinsic value. Competition for habitats, the termination of other life systems in the expression of our interests and purposes, means that many of our valuings involve ending other life systems' valuings, that choosing and valuing intimately involve ending. This threat becomes more pronounced when we move from the killing level (terminating a biological system) to the homicide level (killing a living system taxonomically identifiable as a member of *homo sapiens*). Nevertheless, this emotional threat can only repress, not avoid killing. The extreme case, the murder level (killing a "person") is not avoidable. As I have argued, social existence implies such killing, usually indirect, but not redefined because of that indirection. Such killing requires much stronger justification, of course, precisely because of the increased perception of threat to ourselves. In ethics, the psychological consequences to the actor are also important consequences. They need to be honestly appraised, need to be raised rather than repressed, and the human condition needs to be realistically represented and fully expressed. There was a French film, *We Are All Murderers*, that expressed a significant truth, but only in terms of social conditions as determinants. Our place in existence implies killing in a much broader sense, and the important work of ethics is to accurately determine what killing is necessary and what is unnecessary. Unnecessary pain is a disvalue to an empathic, socialized, interventionist human being.

Accurately looking at the human condition also implies perceptions of worth, of self-image, of self-definition. Our definition of ourselves, like our conceptual categories, cannot be arbitrary or capricious without violating the pragmatic criterion. We are not free-floating creatures in the womb of existence, but part of a larger system and intimately connected in feedback loops to the entire structure. If we choose to define ourselves primarily in terms of reproductive capacity, for example, we not only narrow our potential responses and hence our adaptability, but we incur painful costs that are meaningless because unnecessary. If our sexual identity is narrowed to procreativity, not only have we failed to grasp our extensibility, our creative transaction in experience, but we needlessly accumulate pain for ourselves. These hoary definitions, traditional in the humanities, impose a static, preprogrammed description of the human condition that does not match the data of the life sciences. Within the limits of structural biological predispositions and the limits of socialization (which include both behavior therapy's environmental system perspective and Freudian explanations of bonding and its effects), the human organism is a continuing adaptive process and not frozen in time. Self-definition is not only possible but, if tested continuously in experience, an adaptive desirability, given an affirmative affect and valuation as our context. To a great extent, the humanities have been a force for restricting our human definition, for limiting, by holding previous generalizations of our human possibilities as the only *truly* human ones, for maintaining that human projects must fit within the channels of past experience or they are not human. But the projects we make for ourselves either contribute to our continuing well-functioning in a changing environment or do not, though all the while remaining our human projects. They can be mistaken, they can be pragmatically unjustified, but I fail to see how, conceptually, they could be inhuman.

The genetic enterprise, then, while unfolding potential for change, only reaffirms that the human condition is not static and that we can choose for adaptation. How successfully we make our choices is another story. Failure, systems breakdown, is always the risk of existence. Freezing responses and programs will not eliminate that risk, and may actually increase it. We are all involved in a balancing act, and one of the features of balance is that stopping can bring everything down. Both descriptively in terms of biotic systems, and prescriptively in terms of systems maintenance (survival and well-functioning), the human condition is premised on perpetual adaptations.

Finally, the need for continuing adaptation and definition involves the acknowledgment of stress and anxiety as basic features of the human condition. Can philosophy be of any use in stress situations? Current philosophic tools can be useful in the preliminary stress stage, although if they are viewed as all that is required, they may have more disvalue than value. Rationalization is one of the standard mechanisms of defense and denial. If, however, defense and denial are part of a more complete definition of self, they can supply predictive value, an initial ranking of interests, and the teasing out of alternative choices and consequences. But at this preliminary stage, individuals beginning the choice process are not wholly self-involved. Their full range of interests and needs are only beginning to be existentially threatened. When that does occur, in the late stages of making an important choice, both rational and arational components are present and need to be considered. If a philosophic position assumes concentration only on the rational component and deals with the arational component by attempting to remove it from the decision-making process, such a position will be too narrow to give an adequate and useful analysis of the stress situation, having distorted what is involved. What is needed is a system that, by incorporating factors such as the arational base for intervention, our feelings about pain and suffering, and the presence of a self-affirming or self-destroying affect, can fully explain and thus justify our human choosing and valuing. A decision finally centered on the complete self is thus not an exclusively rational choice.

There is a common confusion I do not want to perpetuate about self-definition, a term sometimes used as a synonym for freedom or free will. My position in no way needs to assume the traditional humanistic notions of free will. These concepts usually wind up unsuccessfully trying to avoid randomness and capriciousness as their real meaning, although I do not see what else "self-willing of an uncaused self removed from the conditioners of experience" could possibly mean. The only other sense is to treat free will as an "as if" description: we like to make choices "as if we had no past learning experiences." If all that this means is that there are self-monitoring, self-correcting programs in the human system, then we are repeating learning theory, and if we wish to label that free will we can, but it will be very confusing in terms of guilt and responsibility. Self-definition, on the contrary, is built up from previous structures and its new configurations are forced (or caused) by feedback loops in the environment. As part of that environment, we change and grow with it. The human condition is not *sui generis*.

The Worth of Human Beings

Medicine, as I have attempted to describe it, is a truly humane science, a notion we pay lip service to, but often do not fully comprehend. Let me describe it in summary:

(1) Intervention as a basic attitude stresses the final importance of human choosing and valuing, rather than the acceptance of nonhuman circumstances. It is an active human attempt to construct human purposes in experience.

(2) Intervention also involves a desire to reduce pain and suffering in other human beings and is grounded on empathy, on socialization. There is behaviorally demonstrated concern and care for other human beings. Their projects are important.

(3) The physician/patient interaction is a human relationship that leads to greater awareness of the patient's interests and, one hopes, to an expansion of the options available. Ideally, it leads to further development of the individuals involved in the process.

(4) Medicine is involved with actual human beings who receive constant feedback from their experiences, and not at all involved with principles of axiological systems devoid of human content. It is empirical, and deeply involved in the human condition.

(5) Medicalization is an attempt to explain human needs and choices in terms other than sin, guilt, and unworthiness.

It reinforces human worth by accurately designating what is and is not under our control, and locating our worth in our purposes and striving for homeostasis, rather than in internal and external events not modifiable by our choices.

(6) By embodying norms of well-functioning, medical science constitutes an affirmation of human existence, but by incorporating the notion of quality of life through well-functioning, becomes more than a survival-at-all-costs position. When intervention is technologically possible, non-intervention then also becomes a human choice. Neither allowing someone to die nor removing someone from existence is equivalent to stating that they have no worth. But the medical enterprise does initially presume that existing human life is potentially worthwhile simply because it is a complex biological structure with mechanisms for maintaining stasis. It is one of our social institutions that really does value human life.

Medicine is certainly a humanism, then, but importantly, its technology, its ability to choose intervention, is an integral part of that humanism and not separate from or in contrast to it. Without the technological developments, the possibilities for choosing intervention are severely restricted. Empathy is not enough, although certainly supportive. Expanding our human choices is also an essential aspect of our humanism.

Future Issues in Medical Ethics

This book has concentrated on current problems that arise in the context of an operating Prenatal Detection Program and has not, therefore, attempted to cover the entire field of medical issues, although I believe it has indicated what are our shared, general concerns, and what is an appropriate theoretical framework. There are some problems, though, that were only hinted at, and I would like briefly to discuss them:

The Demedicalization of the Medical Enterprise

There is an increasing accumulation of intervening variables between the actual provider of the services and the consumer of the services. The proliferation of middle-level planners, regulators, information processors, traffic flow coordinators is fairly well-established. As the delivery system grows larger, these middle levels have also grown larger, which creates an even larger system, which . . . In systems language, this is a positive feedback loop. The cost (not just monetary; for a good review of optimum and maximum concepts, see Gallant and Prothero's *Science* article[107]) is considerable for this middle-level growth. One of the highest costs is that highly trained physicians sometimes do not deliver the service (this is frequently the case in state psychiatric centers). The physician, then, instead of providing the service, coordinates an elaborated system for providing the service. Such bureaucratic growth can only radically alter what we mean by medicine, and because of the positive feedback in such systems, will eventually hit the knee (the point where growth pull buckles under external downward push) of the exponential curve.[108] If what was important to us about medicine was the personal interaction process with an expert and the direct feedback from that experience, such a middle-level growth is not consistent with that interest. In fact, it sets up a system where the information flow from that feedback becomes very problematic.

The Clinical/Academic Conflict of Interest

If academic medicine continues to move into a social concern role, without balancing this with microethics concerns, its goals will begin to conflict more frequently with clinical goals. This has in our American past been a significant problem for medicine. Historically, Flexner was certainly required to redress the balance as a result of Thompsonian pressures, but it is crucial to keep in mind the vital interests of both these forces and the fact that the interests are not always the same. Academic medicine relates to other social institutions in the society as an institution, and how to square that with the individual interests perspective of clinical medicine is not so easy a task. We need to maintain a balance of power if we wish to handle the macroethics/microethics problem.

[107]Jonathan A. Gallant and John W. Prothero, "Weight-Watching at the University: The Consequences of Growth," *Science* **175**, (1972); also, Martin L. Cody, "Optimization in Ecology," *Science* **183**, 1156 (1974).

[108]Roberto Vacca, *The Coming Dark Age*, Anchor Press/Doubleday, Garden City, New York, 1974.

Non-Intervention as
a Required Human Choice

Although a careful and considered choice except in emergency situations, intervention is fairly easy for a physician to choose. Deciding to stop intervening, to make another choice at some step for non-intervention, is more difficult and usually takes an excessively long time. I have case studies from other than medical genetics departments that illustrate this point only too well. The reason is partly lack of ecological sophistication. Once a process is set in motion, we fail to see that allowing it to continue is just as significant a choice (when we have the power to stop it) as not allowing it to continue. We can never do "nothing." But we are also not used to thinking in terms of human choices and control, and vacillate between unrealistic childhood notions of omnipotence or wishes for total dependency. Each new development that gives us the choice of changing circumstances also gives us the choice of deciding in which circumstances to do it and in which not to. If we did not have technologically elaborate newborn intensive care nurseries, many seriously defective babies would quickly die. There would be no human choice. But we have effectively created the choice and to deny it is to artificially restrict ourselves, in a really existential sense. One of the looming issues in medical ethics is our need to come to grips with our responsibility, with our options, rather than trying to disguise events as inexorable.

The Institutionalizing of Caring

As social bonding becomes more complex and as the socialization process breaks down, institutions for replacing these necessary functions are growing. Hospitals, for example, are beginning to create roles for caring for the terminally ill. Social service agencies are examples of what we develop when the possibilities for personal level resources are non-existent. Whether we can adequately substitute such institutionalized for personal caring is a major question that will need consideration as social system complexity increases.

The Benign View of Nature

For a number of reasons—inadequate life sciences education, a wish to ignore the multiple effects of any action, cultural lag in our perception of humanness, unwillingness to accept the requirement of defining ourselves—there is a prevalent view of an ill-defined "natural" as good and of scientific and medical intervention as bad. A sort of noble savage misconception of experience, this trusting Panglossian view can generate much unnecessary pain. It also creates a conceptually untenable separation between science and nature, which is part of the conceptual confusion of this view. My guess is it will become more prevalent as limited resources lead to limited growth and therefore an imposed lowering of expectations. It will be to the social system's advantage, since it will no longer be able to supply all the techical support services or afford to intervene, to have its members believe they can heal themselves more effectively than medical intervention can. There are just enough instances where this will work—because of psychological factors, the self-limiting nature of many illnesses, our ability to put up with pain and discomfort, modern blocks against many disease pathways, effective immune systems—to give credence to this view of nature as benign. Also, the selection process involved in the health of primitive peoples is conveniently forgotten. I expect this view to become more prevalent, however, along with a redefinition of "need," because of this limits-to-growth situation.

Rationing Through Limiting Access

Again because of the likelihood of a limited growth future, access to medical care will become a significant issue, though I doubt that it will be honestly presented as such. Rather, the scenario will probably follow the approach of New York State Health Systems Agencies in limiting availability of CAT scanners, for example, by arbitrarily and nonmedically defining the "need" for them. (The recent Nobel Prize for Medicine award to the developers of the CAT scanner is an ironic comment on that definition.)

Human Self-Definition

A number of medical advances (in genetic intervention, renal dialysis, transplants, life support systems, bioamine systems) are going to force a review of the traditional limits to human definition, limits that are becoming maladaptive. If this is not done, our adjustment to continuing experience can become very inadequate and result in much unnecessary pain. We will need eventually to realize that our continuity to past generations—back even to Olduvai Gorge—comes not from clutching to the very same patterns of response, but rather from our ability to form such patterns in the first place, and that the slow process of evolution makes us related and different at the same time.

Making Natural Experiments Meaningful

I have discussed this previously and only want to restate it as a growing issue. Risk is not avoidable in life, so that any program to eliminate natural experiments would deceive us all. Unnecessary risk is another matter, though its determination is much more complex than the media would have us think. It may be that the Love Canal tragedy fits the category of unnecessary risk, especially in view of the accumulated history of such actions by the chemical company involved. I am not sure that the Three Mile Island incident is so easily categorized. Physicians will increasingly have to determine the consequences of such experiments. To give any meaning to such events, to give sense to individual sacrifice and risk, mechanisms to convert the necessary risks to meaningful, purposeful risks, now lacking, need development. Epidemiology has a long road to go, and without cooperative information transfer, has only taken a small step. How this need to know the effects of natural experiments would effect privacy values is another serious issue.

The Influence of the Media

So far, politicians have primarily been concerned with the filtering of experience through, especially, the eye of the media. All avenues of mass communication can present predetermined categories for the organization of experience and serve as a massive block to correcting feedback from that real experience. How those categories are chosen might prove an uncritical fashion, fad, or hype that, although resembling the fashions in medical ethics, can be much more effective and influential. The process is not clearly understood and self-correcting techniques are not yet well-designated. Since there is a growing concentration in the media on medical topics, how this conveying of experience is accomplished should be a major concern of the medical enterprise and of medical ethics.

Medicine as Mythology

We are not so far removed from the shamans as we would like to imagine. Medicine very directly deals with the experiencing of existential threat, either physiologically because of the serious malfunctioning of our bodies or impending death, or psychologically because of our inability to handle stressful situations or overwhelming affect. Such involvement of our selves is not satisfactorily handled by rational manipulations alone. Metaphorical and mythical elements are not eliminable, nor would I really want them to be. The "secularization" of medicine[109] could be accomplished in this sense only by removing the human elements. To go to a body mechanic model, or a legal contract model, in medicine would leave this large area of human need unmet as well as be bad medical practice, for the reasons I have here outlined. It is the physician's attempt to do something about this existential threat (intervention) that is the mythic requirement and, as well, the ultimate ethical justification of medicine. There is nothing mysterious about this, nor does it create and multiply Occam's unwarranted entities. It is the construction of human meaning in the face of experience that has meaning only through our construction, and is a strong need that, if unmet by one part of the social system, will seek fulfillment in another.

There are risks, of course, to mythic approaches: literalization of metaphor, inadequate safeguards against irrationality, mistaking arationality for irrational license, substituting wishes for reality. To some extent, the mythic aspect of medicine exhibits all of these: that we can fight death and win the fight, that faith in the physician and the treatment will always be rewarded, that things can be made right, that equipment and drugs have magical properties and no unpleasant costs, that intentions to help will result in good consequences, that the physician will save us. Although a risk, however, the elimination of the risk is hardly feasible. Too much is expected of medicine in this mythic role, but then, too much is expected of all human behavior. That lapse into unreality may not be correctable. It can be contained, however, and medicine can fill a mythic need in a reasonable way. Medicine can continue to deal with the total self, with its arational interests and needs, and enhance its completeness as a science. Science without valuing, without arational components, is a meaningless notion. Intervention is importantly arational.

Epilog

Where sympathy is lacking, understanding will not come very easily.

Sigmund Freud, "One of the Difficulties of Psychoanalysis"

[109]Roy Branson, "The Secularization of American Medicine," *Hastings Center Studies*, **1** (2), 17 (1973).

Index of Case Studies

A

Abdominal mass, 23, 155
Abortions, spontaneous, 62, 92, 148,
    184, 185
Achondroplasia, 83, 99
Alcohol abuse, 135
Alphafetoprotein, in research, 131
Amniocentesis, and abortion
    commitment, 14
    indications for, 77
Amniocentesis, risk of, 73
Anencephaly, 61, 154
Anesthesia, risk of, 68
Aniridia, 45
Anxiety, 131
Autism, 51

B

Balanced translocation, 4 and 7: 55
Benzene, 136

C

Caudal regression syndrome, 21
Central nervous system abnormality,
    35
Cerebellar ataxia, 125
Cerebral palsy, ataxic, 50
    with autism, 51

Chelating pump, 117
Chemical dyes, 136
Chloride diarrhea, congenital, 103
Cleft lip/palate, 74
Connective tissue disorder, 44
Consanguinity, 176
Counseling methods, research in, 118
Cystic fibrosis, 91
Cystinosis, 92

D

Developmental delay, 52, 109
Diabetes, 21
    juvenile onset, 67, 126
Dilantin, 77
Dioxin, 135
Down's syndrome, 32, 34, 67, 69,
    84, 103–04, 146, 151, 155, 158
Dwarfism, 36, 83

E

Ectrodactyly, 98, 121
Elavil, 147
Encephalocele, 67, 103, 148
Epidermolysis bullosa, 91

F

Fetal cell sorter, 130
Fetoscopy, 115, 121

G

Gonadal dysgenesis, 66

H

Hearing impairment, 22
Hemophilia, 122, 152
Hernia, diaphragmatic, 68
HLA linkage, 125, 126
Huntington's chorea, 11, 35
Hydrocephaly, 37, 185
Hyperextensible joints, 44
Hyperpigmentation, 54
Hypochondroplasia, 99

I

Infertility, 184

J

Juvenile onset diabetes, 67, 126

K

Kugelberg–Welander syndrome, 180

L

Legg–Perthes syndrome, 91
Lepore anemia, 11
Librium, 135

M

Marfan's syndrome, 66
Maternal age, 62, 63, 69, 110,
    145–148, 151, 152, 169
Meckel's syndrome, 59, 75, 155
Megalencephaly, 37
Menke's syndrome, 183
Mental retardation, 36
Misinformation, 180
Mosaicism, 154, 179
Mucopolysaccharidosis, Hunter's, 90,
    110, 176
Muscular dystrophy, 126
    Duchenne's, 74, 122, 154
Myelomeningocele, 76

N

Neurofibromatosis, 97
Nystagmus, 43

O

Osteogenesis imperfecta, 97

P

Phenobarbital, 152
Phosgene, 136
Porphyria, 96

R

Radiation, 136
Risk figures, 75
Rubella, 69

S

Sex selection, 168, 169
Schizophrenia, 84
Sickle cell anemia, 2
Spina bifida, 74, 99, 131
Spinal muscular atrophy, 179
Sprengel's syndrome, 42

T

Tay-Sachs disease, 89
Thalassemia, 11, 60, 92, 104, 115,
    117, 151
Trisomy 9, mosaic, 154
Trisomy 18: 91
Trisomy 21: 34, 67, 69, 84, 103–104,
    146, 151, 155, 158
Trisomy 21, translocation, 32
Turner's syndrome, 33

U

Upper limb deformity, 42
Uterus, perforated, 181

Index

A

Abortion as intervention, 156–158, 219
Abortion as requirement for amniocentesis, 14–21, 148, 149
Abortion, decision changes about, 152–154
    and homicide, 140, 141, 161–163
    and killing, 138–140, 160, 161, 219, 220
    late, 156
    and murder, 141–145, 163
    psychological costs of, 158–160
Acceptance, attitude of, 26–28
Affective components of ethics, 49, 50, 54, 79, 119, 120, 209–211
Anxiety, analysis of, 64–66, 221
    evaluation of, 58, 59, 70, 71
Aray, Julio, 158, 159
Autonomy, and information, 41, 42
    and responsibility, 108, 109
Ayd, Jr., Frank, 119

B

Baker, Robert, 81
Bayles, Michael, 12, 13
Bazelon, David L., 83
Beck, Lewis White, 197
Brock, Dan W., 206

C

Cassell, Eric J., 10, 54
Categorical imperative, 39
Cavanaugh, James, 17
Chance as fair distribution, 150
Consequentialism, 24–26, 207, 208
Constitutive categories, 190–192
Cost/benefit analysis, 16–21, 207, 208

D

Demedicalization, 223, 224
Denial, direct and indirect, 63
    limits to, 63–66
Deontology, 207, 208
Disease labeling as valuational, 45, 46
Dual roles for physicians, 182, 183, 185–189
Dualism, 132, 133, 208
Dworkin, Ronald, 197, 198

E

Ecological model and consequentialism, 24–26
Engel, George, 14
Expert role, ethics of, 78, 123, 124, 188

F

Fairness, 95, 149–151, 215, 216
Feinberg, Joel, 195, 196
Fletcher, John, 172
Fletcher, Joseph, 50
Frankena, William K., 193

G

Good, instrumental and intrinsic,
    71–73, 163, 216
Gorovitz, Samuel, 52–54

H

Hardin, Garrett, 25, 166, 188, 189
Harsanyi, John, 197, 199, 200, 206
Hart, H. L. A., 196
Hobbes, Thomas, 157, 197, 198
Human condition and ethics, 219–222
Human definition and self-worth,
    220–222, 226
Hypothetical imperatives, 38–40,
    208, 209

I

Individual as object, use of, 28
Information, as important function of
    medicine, 56, 57
    as intervention, 37, 38, 217, 218
Informed consent, 217, 218
Institutionalizing of caring, 225
Intervention, attitude of, 28, 106,
    118, 156–158, 178, 217–219
    choice against, 225
    and compassion, 49, 50
    invasiveness of, 132–135
    necessary or unnecessary, 46–49

J

Justice, 197–200, 215, 216

K

Kantian ethics, 25, 78, 128, 129,
    202, 203, 209
Kass, Leon, 3, 6, 7

L

Ladd, John, 20
Lasch, Christopher, 191, 192
Legalism in ethics, 200, 201
Levi-Strauss, Claude, 127
Limited Growth Model, and medical
    care, 15, 16

M

Macdonald, Margaret, 195
MacIntyre, Alasdair, 52–54
Macklin, Ruth, 197
Macro- and micro-ethics, 165, 167,
    168, 182, 183, 185–189,
    205–207
Maximin principle, critique of,
    198–200
May, Rollo, 191
McCloskey, H. J., 196
Meadows, Donella and Dennis, 25
Media and medicine, 227
Medical ethics, current assumptions
    in, 1, 190
Medicalization, 1, 5, 6, 85
Medicine, academic and clinical, 224
    curative and preventive, 3–6
    error in, 52–54
    future problems of, 223–227
    as generating and imparting infor-
        mation, 56, 57
    as humanism, 222, 223
    as mythology, 227, 228
    proper aim and definition, 3
    real issues in, 58
Moral autonomy, 204, 205
Murphy, Edmond, 111
Murphy, Jeffrie G., 197

N

Natural experiments, ethics of, 137,
    227
Natural law, 127, 128, 193–195
Natural rights, 193–196
Naturalistic ethics, 208, 209
Nature as benign, 225, 226

P

Pain and suffering, attitudes toward, 26–30, 210, 211
Paternalism, 28, 29, 120, 121
Patient, manipulative, 177, 178
  as Open System, whole, 12–14
  as social level term, 101, 141, 142
Physician–patient relationship, 12–14
Piaget, Jean, 46
Possible individuals, possible persons, 81, 82, 100–102, 105
Potentiality, 140, 141
Powledge, Tabitha, 172–174
Pragmatics, ethical system of, 208–211
Prisoners and risk, 124, 125

R

Ramsey, Paul, 118, 119
Rationing medical care, ethics of, 149, 150, 226
Rawls, John, 20, 197–200
Referral, ethics of, 178, 179, 182
Regan, Tom, 197
Rescher, Nicholas, 150
Research, justification for, 115–117, 119, 218, 219
Research, types of, 114, 115
Respect for persons, 78, 128, 129, 202–205
Responsibility, 52–54, 95, 96, 111, 112
  and carriers, 106–109
  and physiological abnormality, 95, 96
Rights model, 78, 106, 127, 128, 193–202
Risk-taking, evaluating decisions in, 81, 123–125, 134, 135
  as gambling behavior, 82, 83
Romano, John, 10

S

Sedgwick, Peter, 45
Self-destruction, ethics of, 78–82, 123, 218
Sex selection, ethics of, 170–175
Sexuality and reproduction, 94, 95
Shainess, Natalie, 159
Sick role and risk, 123–125
Surrogates, ethical justification of, 118, 121, 125, 127–130, 201, 202, 205, 207
Systems theory and "person," 144
Systems theory, feedback loops in, 7–9, 190–192
Systems theory, hierarchy levels and medicine, 9, 10, 138, 213–215
Systems theory, homeostatic concept and medicine, 7–9, 165–167, 170–172, 212, 213, 216
Szasz, Thomas, 45

T

Taylor, Paul, 50, 81
Thomas, Lewis, 56
Thomson, Judith Jarvis, 156

U

Utilitarianism, 78, 79, 128, 129, 205–208

V

Values, precondition for, 39, 40

W

Wild, John, 93, 94
Worth, and carrier status, 93–96
  external, 87, 88
  internal, 88, 89
  of individual, 87–89
  and killing, 101, 102
  and reproduction, 93–96, 100–102, 105